DARWIN'S
APPRENTICE

DARWIN'S APPRENTICE

An Archaeological Biography of John Lubbock

Janet Owen

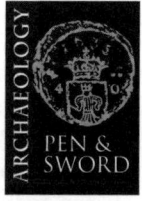

First published in Great Britain by
PEN AND SWORD ARCHAEOLOGY
an imprint of
Pen and Sword Books Ltd
47 Church Street
Barnsley
South Yorkshire S70 2AS

Mammoth image in text:
From the frontispiece of *Prehistoric Times* (Lubbock, 1913).

ISBN 978 1 78159 266 3

Printed and bound in England by
CPI Group (UK) Ltd, Croydon, CR0 4YY

Typeset in Times by CHIC GRAPHICS

Pen & Sword Books Ltd incorporates the imprints of
Pen & Sword Aviation, Pen & Sword Family History, Pen & Sword Maritime,
Pen & Sword Military, Pen & Sword Discovery, Wharncliffe Local History,
Wharncliffe True Crime, Wharncliffe Transport, Pen & Sword Select,
Pen & Sword Military Classics, Leo Cooper, Remember When,
The Praetorian Press, Seaforth Publishing and Frontline Publishing

For a complete list of Pen and Sword titles please contact
Pen and Sword Books Limited
47 Church Street, Barnsley, South Yorkshire, S70 2AS, England
E-mail: enquiries@pen-and-sword.co.uk
Website: www.pen-and-sword.co.uk

Contents

Foreword

Who was John Lubbock? In his own time he was well known. Since his death in 1913, however, all except family and a few of those who share one of his many passions have forgotten him. Now, with the centenary of his death upon us, there seems to have been a resurgence of interest. Two recent full-length biographies have preceded this work and, at his home of High Elms in Kent, there has been a lottery-funded initiative to uncover more about him. In the years since his death at least two short biographical articles have referred to him in their titles as "the Forgotten Man".

To answer the initial question, John Lubbock was a complex man. He has been described as a natural scientist, politician, banker, entomologist, social reformer, anthropologist, archaeologist and many other labels besides. Perhaps this wide variety is one reason behind his obscurity. It has often been said that if he concentrated on fewer subjects, his achievements might have been better known.

John is also my great grandfather. In my family, his name is well known and he stands out amongst a number of high achievers for the sheer breadth of his activity. Once a year, his descendants meet for a family picnic at the former estate of High Elms. The house sadly burnt down in 1967 but there is still much that has been preserved where we celebrate his life and work.

In writing this book, Janet Owen has produced a fascinating new twist on the man. Her achievement is based on many years' hard work assembling facts and making connections that must have been very hard to find in some cases. I have followed the course of this task and the new ideas that have emerged during this time with great interest.

The book concentrates on John's anthropological and archaeological activities. As Janet explains, circumstances caused John and Darwin to be brought together at an early age and the two seemed to "hit it off" right from the beginning. It helped that John had his own inclination towards natural history, a certain enquiring mind, and, later, the time and money to pursue his interests.

DARWIN'S APPRENTICE

Darwin's Apprentice is aptly named. The book takes us through three stages of development of the apprentice. The first is John growing up, learning from the master and watching the development of *On the Origin of Species* (1859). Even in this early period, Darwin was anxious that John should be kept on side as he had begun to value the young man's opinion highly. The second period starts from the publication of that book and chronicles the use of John's collection of objects as supporting evidence for Darwin's theories. *On the Origin of Species* largely left the reader to draw their own conclusions as to the implications of evolution for Mankind. Later works, including those by John and Darwin, took up the challenge of exploring the past of our species. These also required an evidential base, a part of which John attempted to provide. The third stage, during the latter part of John's life was after the case had largely been made. By this time, the contents of the collection served as a more traditional museum collection – mainly referring to things past.

The world-spanning sources of the collection and the variety of routes by which it came together show how Darwin's apprentice learned his lessons well and became a true champion of the cause he espoused.

Lyulph Lubbock, 2012

Preface

*D*arwin's Apprentice tells the story of an important yet forgotten Darwinist, John Lubbock, through the eyes of his prehistoric archaeological and ethnographic collection. Both man and collection are witness to an extraordinary moment in the history of science and archaeology – the emotive scientific, religious and philosophical debate triggered by the publication of Darwin's *On the Origin of Species* in 1859.

This book is a labour of love that began in 1984 when I first became acquainted with John. I volunteered as a teenager at my local museum in Orpington (Bromley Museum) and was asked to unpack and document some stone artefacts wrapped in old sheets of newspaper and stored in dusty brown boxes. As I worked through the collection, I discovered that these objects came from across the globe. Some places were familiar to me and others were a completely alien world. I knew there was a tantalising story that I wanted to uncover about the collector and his collection.

A decade later, I took the chance to study for a research doctorate part-time while I worked at the University of Leicester and later for Nottingham City Museums. I needed a topic that would fascinate me and still hold my interest late at night when burning the midnight oil. The decision was not a difficult one as I recalled the early voluntary work that inspired me to enter into a career in museums. I enrolled on a Ph.D. course at the University of Durham on the 'Collecting Activities of Sir John Lubbock, 1834-1913' (Owen, 2000), and have spent the last fifteen years piecing together the fascinating tale now contained within the covers of *Darwin's Apprentice*. I am delighted that on the centenary anniversary of John's death I have this opportunity to share my enthusiasm for his collection with a wider readership.

Darwin's Apprentice makes use of unique events within the life of the collection to produce new insights into the man. It presents a broadly chronological narrative with every chapter focusing on a particular theme in John's archaeological biography. Each is introduced with a short story inspired by his collection, drawing the reader in to explore a stage in John's life and his involvement in the late nineteenth century turmoil of intellectual debate. After a twenty-five year acquaintance, I feel that I know John well enough to be on first-name terms, and I hope that you will also make a personal connection to his story as you read this book.

John's collection is a wonderful example of the wider storytelling power of historic collections now kept in our museums across Britain. Every object that resides in a display case or sits in a museum store is a window into the lives of our ancestors, and all have rich tales to tell. Who collected me? Why? Where did I come from? How was I used? How did I end up here? Museums today frequently struggle to make use of these old collections from across the globe when they are telling their local story. However, they are important resources that add new and valuable insight into world histories that no documentary evidence on its own can provide.

The final chapter in *Darwin's Apprentice* is written partly in homage to John and his most famous work, *Pre-historic Times*, in which he used his last chapter to express an overtly political view on natural selection and human antiquity. It presents my personal view about the importance of understanding prehistory in modern society, and concern that it is becoming a forgotten part of our past.

Entitled 'Letting the Stones Speak', the last chapter is also an expression of John's personal legacy to me, and one that I hope at least some of you will connect with on completing this book.

Janet Owen, 2012

Acknowledgements

*D*arwin's *Apprentice* is the culmination of an interest in John's collection spanning over twenty-five years. Many people have helped me along this journey and I want to take this opportunity to thank them.

I had a great time researching the collections and associated archives in many wonderful institutions, including the British Museum, Bromley Museum, the British Library, the Royal Botanical Gardens at Kew, the Ashmolean Museum, Bodleian Library, Pitt Rivers Museum, Cambridge University Library, the Danish State Archives, University of Lund, the National Museum of Denmark and the Royal Library, Copenhagen. I am grateful for the kind assistance of staff at all the venues I visited, particularly Dr Alan Tyler (Bromley Museum) who first introduced me to John and his collections; Jill Cook (British Museum) for her expertise and enthusiasm about the project; Alison Roberts and Andrew Sherratt (Ashmolean Museum); Jørgen Jensen and colleagues (National Museum Copenhagen); and Berta Stjernquist (University of Lund).

I would also like to thank all those people who expressed an active interest in my research ideas when I was writing my doctorate; especially my supervisor Professor John Bintliff who stretched my thinking and gave me the confidence to express my ideas; my ex-colleagues and students at Leicester University; Katrina Gjerløff in Copenhagen for her assistance with Danish translation; and generally to all those in Denmark and Sweden for their hospitality and the enjoyable discussions I had with them about Scandinavian archaeology in the nineteenth century.

Thanks also to the University of Leicester and Newnham College, University of Cambridge, for their generous financial assistance in support of the research element of my Ph.D.

More recently, I am very grateful to a number of people who have encouraged me to continue with my research interest, and to make wider connections between John's story and the histories of archaeology, science and empire. These include my ex-colleagues at the National Maritime Museum; Arthur MacGregor who gave me the honour of contributing a chapter to his centenary anniversary volume commemorating Sir John Evans; and Clive Gamble whose kind encouragement has given me the confidence to believe in the wider value of my work. My thanks also go to both Arthur and Clive for their comments on earlier drafts of this publication.

Finally, I want to thank those people who have helped me throughout this long journey. Lyulph Lubbock has always been a generous source of information and advice, as well as giving me ready access to, and permission to use, the various family archives and information sources concerning John and his collections. I am greatly privileged that he has written the 'Foreword' for this book.

A big thank you also goes to my family and friends for their support, especially to my husband for all his enthusiasm, perseverance and technical input; to my mother for her invaluable research assistance; to all the members of my family, especially Lauren and Mikey, who accompanied me on a trip to Avebury one Sunday in 2010 and filled in the activity sheets I had given them; and to my sisters who read several drafts to make sure the book was telling a good story!

This work is dedicated to my mother and father who have inspired me so much in my life.

Introduction

Laying out the Collection

'Had some of the Down & Farnboro people to see the Museum. They had tea in the dining room & we put out microscopes, photographs, savage implements & c. all around the billiard room which made quite a bright & interesting show. They all seemed very pleased. I spoke for about 1/2 an hour.'[1]

The billiard table had taken on quite a different function from usual. The blue cloth was stretched carefully over the green baize to create a large makeshift display table, with four finely carved mahogany legs protruding from under the fabric. The normally quiet and tranquil solitude of this country house retreat was about to be invaded by an audience of curious local residents from Downe[2] and Farnborough, intrigued by the prospect of seeing Sir John Lubbock's extraordinary collection of prehistoric archaeological and ethnographic artefacts.

In a contemplative mood, John picked through the vast range of stone and bone implements in his collection. Every few minutes or so he selected one or two items which he placed carefully on the blue cloth. This was a task he loved – touching the objects, in awe of their great age, and remembering the stories they whispered to him from years long past.

Occasionally, he held a piece up to the light to examine the detail of its workmanship, or to read an inscription he had written on the item many years before. Perched on the lamp-table beside him was an open notebook, entitled 'Catalogue of My Collection'. It contained handwritten details of almost every item John had acquired over the last thirty years. John always made sure to check his facts before discussing an item in public, although he could often recall what an item was and how he acquired it. In his head he rehearsed the stories he would use today to bring these items alive for his audience as he placed each object in sequence.

His mind was distracted momentarily as he heard a loud clatter of china from the dining room. Alice's calm voice, orchestrating the preparation of tea for their visitors, put his mind at rest that all was under control. Alice Pitt Rivers and John occasionally threw open the doors of their private estate to invited local people during the seven years Alice lived at High Elms with John since their marriage. They would have tea served in Spode cups,[3] study the curious artefacts and listen to John talk about human evolution and his late friend, Charles Darwin. At first he found the most difficult part of these events was choosing which of the 1,000 plus items in his collection that he should present and talk about. This job had been made slightly easier after 1890 when, with the assistance of Augustus Wollaston Franks and Mr Oldland from the British Museum, he created a display of books, flint implements and other ethnographic items in the hall at High Elms.[4] Now at least any visitor could easily admire key items in his museum, and others could be brought out for display according to the occasion.

Today's audience would, as ever, respond in various ways to his ideas. Several might have visited High Elms on previous occasions and have some understanding of evolution and the idea of human progress, but John still expected some critical questions concerning the relationship between these theories and his Christian faith. For others, this would be a first visit and the opportunity to see something of the inner workings of High Elms. He knew from experience that several welcomed the opportunity to handle these antique artefacts. It would be a tiring day for both Alice and John but a rewarding opportunity to educate local working people about science, evolutionary theory and ultimately the liberal values of moral, social and economic progress he so strongly believed in.

After another half-hour the display was ready. John looked up from his deliberations and admired his handiwork. A good representation of his entire collection lay before him, complementing the items already exhibited in his hall museum. He planned what to say about each object as hidden memories came flooding back. This collection, in so many ways, captured the drama and debate of those exciting years during the 1860s when Darwin's theories surrounding the origin of species challenged religious belief and the accepted norms of the scientific community.

As John surveyed his latest display, his gaze turned to the selection of fishing implements he had placed on the table and rested on a bone harpoon with the number '277' inscribed on it, an object very close to his heart.

Chapter 1

The Bone Harpoon

'When I am a man I mean to travel all over the world, and find out everything, every plant and animal, and what lies at the bottom of the sea; and I mean to go to China.'

(John, aged six years, as recounted in the diary of his mother, Harriet Lubbock. Grant Duff, 1924, p.7)

John would always remember the story behind catalogue number '277'; a long bone harpoon serrated along one edge. His master and friend, Charles Darwin, had given it to him in 1864. A moment of sadness passed a shadow across his mind as he recalled the day ten years ago when his shoulder had slowly carried the weight of this great man's coffin up the long aisle of Westminster Abbey. Then he smiled as he remembered Darwin recount the stories of his expedition to Tierra del Fuego on HMS *Beagle* during which Darwin had been given at least two such harpoons by local inhabitants. How fascinated and inspired he had been as a teenager to hear these stories. How much he owed to that wonderfully bright and intelligent man who had taken him under his wing and mentored him throughout his life.

'A talk with him was as good as sea air...I am one of the thousands whom Darwin has inspired by his writings, and of the few still living who have had the inestimable privilege of his friendship.'

(Grant Duff, 1924, p.26)

He had been thirty years old when 'Mr Darwin' presented him with the harpoon numbered '277' as a gift in the 1860s.[5] However, at this particular moment he remembered the time many years before when he was first captivated by Darwin's remarkable tales of the HMS *Beagle* voyage (Plate I). They both escaped into stories of the alien, inhospitable and wild lands of Tierra del Fuego as they walked along the Sandwalk (Plate V) at Down House with the gravel crunching beneath their shoes. Darwin would talk of the mountains, ice, forests and ravines, and of the stormy seas that lashed HMS *Beagle* and sought to sink her. He would relay tales of encounter with people whom he saw to be so primitive. He would describe his scientific observations and the myriad of questions these threw up in his mind. Why had he found marine shells in the rocks at the summit of Mount Tarn, 2600 feet above sea level? Why did people choose to live in such a difficult environment? How could the Fuegian people seem so different to the human species as Darwin knew it, yet Jemmy Button had demonstrated that these people could act in a civilized manner with appropriate education?

Jemmy Button was undoubtedly a particular topic of fascination. He was one of four Fuegians that the captain of HMS *Beagle*, Captain Robert Fitzroy, brought home to England on a previous voyage in 1831. His idea was to teach them English, Christianity and the ways of western civilization before taking them back to Tierra del Fuego as missionaries. Only three of the Fuegians survived the trip to England and were given the names: Jemmy Button, York Minster and Fuegia Basket (an eleven year-old girl). A primary purpose of the HMS *Beagle* voyage was to return the three Fuegians and set up a mission station at Woollya Cove in Ponsonby Sound – the original home of Jemmy Button. In the close quarters of the ship on a long ocean voyage, Darwin had plenty of opportunity to observe and strike up a relationship with Jemmy and the others.

Darwin was present when HMS *Beagle* left the three Fuegians at Woollya Cove with the missionary, Matthews, to settle into newly built huts and freshly planted gardens with all the trappings of western civilization – wine glasses, tea trays, fine white linen and decent clothing (Desmond & Moore, 1991, p.134). Darwin was also there when the crew of HMS *Beagle* returned several weeks later to find the mission systematically looted and Matthews wishing to return to the ship. When HMS *Beagle* returned again one year later they found the huts deserted, the gardens overgrown and Jemmy Button clothed only with a blanket round his waist. Over dinner at the Captain's table, Jemmy used cutlery properly, spoke English and told of what had happened. York Minster and Fuegia had left him, taking his possessions and clothes, and Jemmy had returned to Fuegian society where he was happy with a new wife and plenty of food (Desmond & Moore, 1991, p.147-8).

John remembered all of these stories and how he was enchanted by the passion and intensity with which Darwin spoke about these things. He also remembered how Jemmy Button had given Darwin two 'spearheads' made by himself as a parting gift at that final meeting (Desmond & Moore, 1991, p.148; Fitzroy, 1839, p.327). All of these memories came flooding back as he gently cradled the harpoon numbered '277' in his hand. What an important gift this had been from his Master in 1864 – a true mark of a friendship nurtured over many years. John may even have used this artefact to illustrate his own ideas on human evolution in *Pre-historic Times*[6] (Plate II).

Darwin's Apprentice

In the early years of their acquaintance John had little idea of the lifelong significance his relationship with Darwin would have. Nor did he have an inkling of how Down House was to become the epicentre of a scientific revolution with global importance. How would the events of the next seventy years have been different if the Lubbock family had not made High Elms their main home during the 1830s, and had the Darwins not moved in to the neighbouring property of Down House in 1842? Both families wanted to escape the claustrophobia and intensity of London living, and were attracted to the idea of raising a young family in the peaceful North Kent countryside. The head of the Lubbock household, John William Lubbock, and the head of the Darwin household, Charles Darwin, were both intellectual thinkers and wanted the space to reflect on their work. As neighbours, they found they had shared interests, political values and a connected network of friends.

Only five years before his 100-guinea horse and carriage drove up to Down House for the first time, Darwin had stepped ashore at Falmouth after a voyage around the world. Shortly after completing his scientific studies at the University of Cambridge, Darwin had taken up a once in a lifetime opportunity to accompany Fitzroy on his survey of the coasts of South America. This voyage on the HMS *Beagle*, which began as a two or three year expedition at the end of 1831, turned into a full circumnavigation of the globe lasting five years. During that time, Darwin put his naturalist skills and interests into practice, collecting thousands of observations, notes and specimens from across the continents with which he returned to England. He immediately set about writing up his travels, studying his collections and sorting out his personal life. In early 1839 he married his cousin, Emma Wedgewood, and moved to London from the family home in Shrewsbury. By the time the couple arrived in Downe they were accompanied by a son, aged three, and a one-year old daughter. Emma gave birth to a second daughter within days of arriving at Down House but sadly Mary Eleanor died a few weeks later. Their new life was tinged with tragedy from the start, but Emma and Darwin intended to stay at Down long-term. They brought with them their family, notebooks and collections from HMS *Beagle* and unfinished business on the subject of evolution (Desmond & Moore, 1991, p.301-4; Burkhardt, 2008, p.xxx).

The large estate and mansion called High Elms (Plate VI) was a couple of miles down the road, with its grounds adjoining the Down House property. The owner, John William Lubbock, was a successful partner in the family bank, Lubbock & Company, which was established in 1772 and located in the City of London at 11 Mansion Street. The company and family had made their fortune over the last fifty years from the great economic expansion associated with the industrial revolution and the building of the empire. However, in his heart J.W. Lubbock was a devoted astronomer and mathematician. Although he published little of major scientific impact,[7] he was awarded the Royal Society gold medal in 1833-4 for his work on tides and tidal tables and was highly influential in intellectual London circles. He had two terms of office as the Treasurer and Vice-President of the Royal Society during the years 1830-46. Educated at Eton and Trinity College Cambridge, he was also the first Vice-Chancellor of the University of London (1837-42) and believed in the importance of opening up educational and scientific opportunities to people of non-Oxbridge and non-Anglican background. He married an energetic and 'liberal minded' Harriet Hotham in 1833.

Into this wealthy family, their first son, John, was born on 30 April 1834 at 29 Eaton Place, London. Curious and questioning from a young age, he appears to have had an interest in observing the world around him even at the age of four:

'His great delight is in Insects. Butterflies, Caterpillars or Beetles are great treasures, and he is watching a large spider outside my window most anxiously.'
(Entry in the diary of John's mother, Harriet, 1838. Hutchinson, 1914, p.6)

However, to the disappointment of both John and his father, these scientific interests were not actively catered for at Abingdon Abbey and Eton College, which John attended. This fact, combined with concerns regarding the young boy's health, contributed towards a decision in late 1848 to remove him from Eton College and start him in the family business helping his father at fourteen years old.

He travelled to Sydenham with his father by horse and carriage, where they both caught the train up to town. John missed having regular contact with friends and colleagues of his own age and on occasion talked of feeling lonely. However, his diaries show how he quickly filled his days and nights with structured timetables of study, work and social entertainment. He continued his scientific and mathematical education with his father, although J.W. Lubbock had little patience as a tutor and thought less highly of natural history studies. John would read seven or eight hours a day, including on the train into work, especially about the subjects of biology and geology. His timetable for study on days when he was not in the City began at 6.30am and finished at 12 midnight. He would spend set amounts of time on mathematics, natural history, poetry, political economy, history, sermons and German. His natural history studies included reading and work with the microscope that Darwin had encouraged his father to buy for him.

This studious work was encouraged by his parents but also demonstrated a significant self-determination and interest from John himself. He certainly had no social or economic need to drive himself so hard. There is no doubt that Darwin became a highly influential figure in motivating John to commit so much of his time and energy into his scientific and wider studies.

'He induced my father to give me a microscope, he let me do drawings for some of his books, and I greatly enjoyed my talks and walks with him. My first scientific original work was on some of his collections, and appeared in the Natural History Magazine for January 1853. In 1849 I was elected a member of the Royal Institution, and in 1853 attended my first meeting of the British Association. In 1854, I was introduced to Sir C. Lyell and Sir Joseph Hooker, in 1855 to Kingsley, Prestwich and Sir John Evans, and joined the Geological Society. In 1856, I met George Busk, Huxley and Tyndall, and the following year was elected a member of the Royal Society.'

(Notes by John reproduced in Hutchinson, 1914, p.23-4)

The paucity of documentary evidence pre-1852 makes it difficult to confirm exactly when the relationship between Darwin and John truly blossomed.[8] However, it was probably strengthened after John left Eton College and started working for his father at the bank in 1848. Darwin was thirty-nine years old and John twenty-five years his junior, but both shared a social loneliness at this time that could be partly overcome by their mutual curiosity in, and love of, natural history.

Since returning from the HMS *Beagle* journey in 1836, Darwin had focused his energies on researching and publishing the many specimens and observations he had collected during his travels. Following the publication of *Geological Observations on South America* in 1846, the only specimen left to publish was a single barnacle species collected from the shores of southern Chile in 1835. The other piece of work awaiting completion was his note on species and transmutation. He had started thinking about this subject in 1842, knowing from the outset that his ideas were highly controversial and would possibly ostracize him from his friends and wider society. Transmutation – the belief that all animals were descended from a common stock and changed through variation – was a scientific idea that at this time was closely associated with the very real threat of social unrest and revolution. Darwin's ideas became a secret that he divulged cautiously to a very few confidantes over the next thirteen years for fear of repercussion.

During the 1840s he probably shared his thoughts with only one other person, Joseph Dalton Hooker, a naturalist who Darwin recognized as a kindred spirit (Desmond & Moore, 1991). Ever since Hooker returned from his epic voyage with Captain James Clark Ross to chart the Antarctic continent in 1843, Darwin had carefully cultivated this young man and nurtured him as a sounding board for his radical ideas whilst also encouraging Hooker to research relevant questions when studying specimens. By 1847, Hooker was spending weekends and weeks at Down, and in January Darwin shared his draft manuscript on species and transmutation with Hooker for the first time. Hooker did not necessarily agree with Darwin but provided a calm and measured opinion, which was what he required. However, Hooker also had his own shock for Darwin in January 1847 when he told his mentor that he planned to set off on another expedition. In November that year, he sailed for the Himalayas with instructions to collect plants for the Royal Botanical Gardens at Kew in London. Although he kept in regular correspondence, he did not return from this trip until 1851. To cap it all, Darwin's father was dying and his own personal health was also deteriorating (Desmond & Moore, 1991).

Darwin was at a crisis moment both in his life and studies. At the same time the eldest son of his neighbour arrived back at High Elms from boarding school with a keen appetite to learn more.

'He felt unable to see anybody, except perhaps Sir John Lubbock's teenage son John. The boy's fascination for his microscope helped rescue Charles from total despair. He escaped with the lad into an unimaginable microscopic world, teeming with life. He became John's scientific father, in a way, obtaining for him an identical instrument. He was carrying on the Doctor's [his father's] practice – giving the wise guidance that his father had given him.'

(Desmond & Moore, 1991, p.361)

In the same way Darwin had been his father's medical apprentice, practicing on the sick poor of Shropshire, at the age of sixteen (Desmond & Moore, 1991), John quickly became Darwin's scientific apprentice. Perhaps this relationship was driven purely by psychological and social need. However, it is not beyond the bounds of possibility that Darwin also saw in John someone from within the social elite, like Hooker, who might be groomed to become an independent champion for transmutation.

John was an open book to Darwin's exciting worldview and his influence is reflected in the way John developed interests in geology, botany, zoology and entomology over the next few years: the very areas of significance to a debate regarding transmutation. He learnt method and observation from Darwin and the skills of microscopic dissection. In 1848 he began the *General Notebook of Natural History* (now housed at Down House) in which he detailed various observations and drew sketches. On page 93, he recorded a brief discussion about the races of man.[9] In 1849, he joined his first learned society, the Royal Institution, and in 1850 gave his first public lecture, on the wireworm, to an audience at Downe. During that year, Darwin successfully proposed the sixteen year old for membership to the Entomological Society.[10] Furthermore, in 1851 Darwin requested that George Newport lend him and John an old pair of dissecting scissors because he 'has taken a passion for dissecting.'[11]

Darwin was beginning to act as John's intellectual patron within the scientific establishment:

'I am much obliged for your kind message to Mr. Lubbock, which I will not fail to give: he is a remarkably amiable pleasant young man, and he has a strong taste for dissecting insects. A little conversation with you, would, I think be of great service to him.'[12]

When John turned eighteen in 1852, Darwin commented on how he was developing into an 'excellent naturalist' (Desmond & Moore, 1991, p.400). He started studying natural history specimens at the British Museum (Department of Natural History) that had originally been collected by Darwin on the voyage of HMS *Beagle*. In this way, Darwin began to introduce John to the idea of collecting specimens. Darwin had an interest in collecting from an early age, and his global trip had provided a dream opportunity to acquire all sorts of material: plants, fossils, shells, rock samples, insects, birds and other animals. He sent some back to his tutor, Professor Henslow, at Cambridge University during the voyage while others were collected from the quayside in Woolwich, HMS *Beagle's* final berth. These were distributed to a wide range of specialists for study and preservation, including the British Museum (Department of Natural History). John would have learnt about the process of collecting and recording both through his work at the Museum and through conversation with Darwin about the specimens he kept at Down House (Desmond & Moore, 1991, p.204; Burkhardt, 2008, p.xx-xxi; Healey, 2001, p.42).

In the early 1850s a new dawn was breaking in wider intellectual society led by John Chapman, Herbert Spencer, George Eliot, John Stuart Mill and Thomas Huxley. With the *Westminster Review* as their platform, these members of the establishment were championing the legitimacy of change and evolution without revolution. The work of this group, and the success of his latest barnacles publications (Darwin, 1851-54), appears to have filled Darwin with a new confidence in his ideas about transmutation. He met Huxley at the Geological Society in April 1853 and received a Royal Society Medal in November of that year in recognition of his barnacles work. He may also have started sharing his radical ideas with his young protégé, John, who came of age both literally and scientifically during the period 1852-1855.

In January 1853, John published his first scientific papers (Lubbock, 1853a; 1853b). On 22 February 1853, he went to 'Mr Darwin for a little advice'.[13] Perhaps the advice he sought related to a particular piece of scientific research he was undertaking. Equally, he might have been interested in how he could become more involved in the scientific community and take forward his own personal ambitions. He would already have been taught by Darwin to take every opportunity to observe the natural world around him. His February walk through the woodland and open country from High Elms to Down House would undoubtedly have been one such moment. He would know about the chalk geology from which the rolling North Downs landscape had been sculpted over millions of years. As he mulled over in his mind the advice he was seeking from his master, he might have stopped to admire the array of fungi growing on tree branches, or the large flint nodules with their gnarled and contorted shapes exposed by the roots of a fallen tree. He would have felt the chill breeze on his face as he reached the top of the rise and left the protective shelter of the valley to walk across the fields towards the tiled rooftops of Downe village. On his arrival at Down House, he was sure to have been met with a warm welcome before asking Darwin for the advice, on whatever it was, that he was seeking.

In September 1853 John attended his first British Association for the Advancement of Science (BAAS) meeting in Hull. He called on Darwin again on 18 August before heading off for Hull on 7 September. We also know that on 22 September:

'When the ladies were gone to bed Charles Henry Strickland and I had a discussion on the mutability of species, and I believe we all agreed that it seemed probable that they might change into one another.'[14]

It is hard to imagine that John and Darwin were not sharing their thoughts on transmutation at this stage. After the BAAS meeting, John spent time travelling through the Midlands and the North, returning to London by way of Cambridge where he met up with Charles Kingsley. He returned home at the end of November and immediately began 'sorting out the freshwater species of Mr Darwin's South American Crustaceans.' On 15 December, he reported that he had begun to examine them.[15]

Darwin's role as patron continued:

'You will have seen my friend and neighbour, Mr Lubbock, has been working a little on the lower Crustacea: he is a remarkably nice young man, only a little above 18 years old: - if you can ever give him a little encouragement it would really be a good service, for he has great zeal, and for so young I should hope, has done well; and if he can resist his future of great wealth, business and rank, may do good work in Natural History.'[16]

By the age of twenty, John was helping Darwin to study his Southern Atlantic crustacean specimens from the HMS *Beagle* voyage. He was also comparing them with material acquired by Dr Sutherland during a recent expedition to the Arctic region, 1850-1, and from the Cape of Good Hope.

He began to undertake original research and contributed actively to wider scientific debate within learned societies. In August 1854, he published 'On some Arctic species of Calanidae' in *Annals of Natural History* (Lubbock, 1854). On receipt of his complimentary copies, he immediately sent a copy to Darwin, Richard Owen and Kingsley, as well as to friends and family. John then attended the Entomological Society and Royal Institution meetings and Darwin invited him to dine with Sir Charles and Lady Lyell and Dr and Mrs Hooker on 27 October 1854.[17]

The opportunity to meet and converse at dinner with these icons of science must have been an exciting opportunity for John. We can imagine the anticipation with which he dressed for a dinner at Down with Hooker, the eminent and globetrotting botanist, and Lyell, the internationally renowned and forward-thinking geologist who was Darwin's great friend and scientific inspiration. There would have been a multitude of thoughts and questions going through his mind as his carriage drove the two miles from door to door. He would have known that this was a special meeting for Darwin as well. John would have sensed new energy and purpose in his master's approach to life over the last few years. He saw how Darwin was reengaging with the wider intellectual establishment, participating in the Royal Society Philosophical Club and the Linnean Society. In September Darwin's second volume of research into barnacles had been published (Darwin, 1851-4) and John had helped prepare the plates. He would have felt too, perhaps, the hope that Darwin now held in the young and emerging scientific generation who might be receptive to ideas on transmutation. For example, John Tyndall was Chair of Natural Philosophy at the Royal Institute, Huxley was now a teacher at the Royal School of Mines and ran the science section of the *Westminster Review*, and Hooker was soon to become Director of the Royal Botanical Gardens at Kew. Darwin had invited Lyell and Hooker to Down House for a few days in October 1854, taking this opportunity to explore their views on species and transmutation (Desmond & Moore, 1991, p.415).

Joseph Parslow, Darwin's butler, and his small household team, would have prepared and served dinner on 27 October. Emma Darwin, as hostess, would have worked hard to ensure that Lyell's quiet tone and presence did not flatten proceedings (Healey, 2001, p.167). Undoubtedly, any in depth scientific conversation that evening would have been reserved for once the ladies had retired, when John would have sat and listened to Darwin, Hooker and Lyell. Over the period of Lyell and Hookers' stay at Down, Darwin concluded that Hooker was beginning to come round to the idea of transmutation. Lyell, however, was nervous and uncertain (Desmond & Moore, 1991, p.415). The opportunity for John to be part of this debate was undoubtedly important to him, although he may not have realized its true significance at the time. Other things, of more immediate interest, were also taking place during 1854 that would have an impact on his life story. Ellen (Nelly) Hordern, his future wife, started to spend time at High Elms and, on 30 December, at the turn of his twenty-first year, John was made a partner of Lubbock & Company.

In 1855, John was appointed to the Council of the Entomological Society and began his duties on 5 February.[18] He continued to work on Sutherland's Atlantic Calanidae and South American Entomostraca collections, fitting his research into the day early in the morning, late at night and whenever else he could (including dissecting 'a new Pontella after church').[19] He generally spent two nights a week in London attending the Entomological Society, Royal Society, Royal Institute, Microscopical Society and Geological Society meetings. He continued to visit Darwin – for example he rode over to Down House on 5 September of that year.[20] John could not have helped but notice the scientific experiments Darwin had started conducting to test his ideas about the mechanics of natural selection. He would have spotted the plethora of small dishes crammed on the mantelpiece in his master's study and smelt the aroma of boiling carcasses immediately as he walked through the door. Darwin was working on two main experiments during 1855-56. The first involved the germinating of seeds in sea-salted water to demonstrate that seeds could be carried to new locations many miles away from their geographic origins by the sea. The second experiment immersed Darwin in the world of pigeon fanciers and in the collection of pigeons. He used this particular example of domestication to examine the role of fanciers in selecting minute variations between birds, and the sculpting of these into significant changes over time. He built a pigeon house in the garden for fantails and pouters. He killed them with potassium cyanide and boiled their bodies to expose the bones for study. He collected specimens from a wide variety of local, national and international sources, including High Elms.[21] He drank with breeders and joined their clubs. For a young man of twenty-one years, these experiments were a master class in the importance of collecting specimens, observation and evidence gathering when testing a hypothesis.

On 30 June, John made his first nationally significant discovery that started to make his name in the scientific community. A few days earlier, he had visited gravel pits in the Croydon and North Kent area with the geologists Lyell and Joseph Prestwich,[22] having probably met Prestwich at the Geological Society earlier that year. On the last day of June, John went with Kingsley to explore the drift geology of Maidenhead. During their fieldwork, they discovered part of an unusual animal skull. Excited by their find, it was taken for identification to Owen and others at the British Museum (Department of Natural History):

'I had a note from Lyell this morning in which he says you have found the first Ovibos moschatus [musk ox] ever discovered fossil in England. I must congratulate you on such a capital discovery. Considering the habit of ovibos, and the nature of drift-beds, I declare, I think it is one of the most interesting discoveries in fossils made for some years. I congratulate you and may this be the first of many interesting geological observations. Yours very truly C. Darwin. I wish you could have come here on Tuesday. Adios.'[23]

Darwin, Lyell and Owen were excited because John and his friend, Kingsley, had found a fragmentary fossil of an animal in Maidenhead that in the nineteenth century, and today, only lives in the Arctic regions. Such a discovery raised significant questions for geology and religion, helping to challenge the Christian-based Ussher chronology, used by western society, that dictated the earth had been created in 4004 BC – less than 6000 years ago. John wrote to Lyell asking if he would propose him as a member of the Geological Society, to which he was elected in December 1855.[24] By this time he was also spending his last few months as a bachelor. On 13 October he had written to Nelly asking her to marry him and on 20 October she had replied to say 'yes'.

John married Ellen Frances Hordern at the picturesque St. Mary's Church in the small village of Rosthern, Cheshire on 10 April 1856 (Plate IV). As they walked out of the red sandstone church as husband and wife and passed through the seventeenth century wooden lych gate, they began a new chapter in their young lives. By the 24 April they were ensconced in their new married 'quarters' at High Elms and it was not long before Nelly would realize the influence of Darwin living next door. On 24 April, Darwin wrote to John:

'Before I come to the purpose of my note, let me give you my very sincere congratulations on your marriage; & I do not believe that any of your many friends can send you more cordial & true good wishes.
    Can you get "leave of absence" & come & dine here on Saturday

at 7 o'clock to meet Dr & Mrs Hooker, Mr Wollaston, Mr & Mrs Huxley.

Mrs Darwin had intended (but is not quite well at present) to have ventured on a very early call on Mrs Lubbock in hopes of persuading her to come with you, but I fear there would have been but very little chance of the party at High Elms letting her leave home so soon.'[25]

John could not have refused such an invitation to meet with Huxley on his first trip to Down House. This dinner was yet again more than a mere social occasion. It was a contrived opportunity for Darwin to engage with Hooker, Frederick Wollaston Hutton and Huxley, and to determine their views on evolution. He was particularly concerned to bring Huxley on side. He could see the potential of his intellectual power and was keen to ensure that Huxley did not set himself against evolutionary theory before Darwin had a chance to publish his ideas. Darwin was already aware that Huxley would make a dangerous opponent. Scientific subjects were undoubtedly part of the topic of conversation served up with dinner on the evening of 26 April, but the focus of Darwin's interrogation was left until the next morning, after John had returned to High Elms. After breakfast, Darwin invited Hooker, Huxley and Wollaston into his study for individual 'interviews'. He had a pile of slips for each with different questions that he wanted them to answer. By the end of the weekend, he had concluded that Hooker was definitely onside, but Huxley had objections to the theory of transmutation. Darwin wrote these objections down and one by one prepared counter arguments. He was also concerned about Huxley's passionate 'hot' temper (Desmond & Moore, 1991, p.434-7).

Despite this note of caution, Hooker, Huxley and Tyndall were becoming close friends for Darwin in influential places. The three of them were keen to see science develop as a profession and all shared a love of the new extreme sport, mountain climbing. Huxley was appointed as Fullerian Professor at the Royal Institute and became an examiner at the University of London while Darwin and the Lubbocks pulled strings behind the scenes to assist. On 27 May, Darwin wrote to Sir John William Lubbock, as Chancellor of the University, providing a reference for Huxley's application for the Examinership in Physiology and Comparative Anatomy. He backed up his word by referring to John, saying 'I think if you will ask your Son he will agree with me in this.'[26] Sir J.W. Lubbock used his influence at the University to assist in Huxley being offered the post.

John continued his research activities throughout 1856 and into 1857. He had published his first papers on the freshwater entomostraca of South

America in the *Transactions of the Entomological Society* (Lubbock, 1855a; Lubbock 1856). He continued his work on the specimens collected by Dr Sutherland, and started investigating the methods of reproduction in Daphnia ('water fleas'). Darwin supplied him with references and promoted his work to a wider network of scientific contacts. He commented on John's draft paper concerning Daphnia, answered questions and recommended that Huxley also be asked to review and comment.[27] He invited John to dine at Down to meet Professor Henslow.[28] He seconded John's nomination to become a member of the Athenaeum Club in March 1857. He also praised John for his considerable success in publishing the Daphnia paper (Lubbock, 1857a) in the *Proceedings of the Royal Society*:

'At the Philosoph. Club last Thursday I overheard Dr Sharpey speaking to Huxley in such high & warm praise of your paper & Huxley answering in same tone that it did me good to hear it. And I thought I would tell you, for if you still wish to join the Royal Socy., I shd. think (Sharpey being influential in Council & Secretary) there cd be no doubt of your admission. Even if you were not admitted the first year it cannot be thought the least disgraceful. I am not aware but perhaps you have already been proposed.'[29]

Darwin, at the prompting of Lyell, had begun in earnest to write up his own ideas on natural selection for publication. He started the chapters on breeding and artificial selection, domestication, distribution, fertility and sterility, the struggle for existence, natural selection and variation. He sent selected chapters to Hooker and Huxley for their comment and gained strength from their response. His confidence was buoyed by Huxley's praise for Darwin's barnacle books. However, Huxley still did not understand, nor did Darwin fully share with him, his views on natural selection (Desmond & Moore, 1991, p.459). Darwin was becoming increasingly nervous about the row brewing regarding human origins. Owen, perhaps sensing that transmutation was about to emerge as an issue, started to build his case for the opposition: for the influence of god in human design. He argued that the brains of apes were clearly different to those of humans because of the presence of a unique lobe, the hippocampus minor, and the fact that their cerebral hemispheres were larger than any other mammal. They were a subclass set apart and there was no continuity, no case for mutation. Darwin fell ill again, perhaps in part due to the internal conflict and stress generated by his ideas on transmutation, and went away for spa treatment.

John did not help Darwin's condition when he spotted a mistake that Darwin had made when tabulating the ratio of varieties in plants.

'You have done me the greatest possible service in helping me to clarify my Brains...I have divided N. Zealand Flora as you suggested...I am quite convinced yours is the right way; I had thought of it, but shd never have done it, had it not been for my most fortunate conversation with you.

'I am quite shocked to find how easily I am muddled, for I had before thought over the subject much, & concluded my way was fair... What a disgraceful blunder you have saved me from...It will take me several weeks to go over all my materials. But oh if you knew how thankful I am to you.'[30]

This letter suggests that John was involved in conversations with Darwin regarding the content of his evolution and transmutation work. In a small way, he was also becoming involved in the Huxley, Hooker and Tyndall network. He was a member of the Linnean Society that Hooker, Tyndall and Huxley had set themselves the task of re-energizing. As a member of the Athenaeum he may have assisted in getting Huxley elected in January 1858.

As well as his scientific work during 1857, which included studying the muscular system of larvae, the year was also busy for John on the work and domestic front. He fitted his scientific work in and around an increasingly central role at the bank as partner and around the responsibilities of a family man following the birth of his daughter, Amy, in March 1857. In this year, he was also hard at work helping to establish a new system for clearing cheques between London and country banks.

It was 1858 when everything moved up a gear in the debate on evolution generally, aided by John's specific contribution. In March, Huxley attacked Owen's work on alternative theories. He challenged the idea of human beings existing in a separate sub class and argued for anatomical continuity between humans and apes. He also published a detailed anatomical study of embryogenesis in aphid insects (Huxley, 1858), which had been read at a meeting of the Linnean Society on 21 January 1858. This challenged Owen's explanation of parthenogenesis.[31] At this time, John was working on the development of eggs in the related Coccus insect family[32] and appears to have sent, through Darwin and Busk,[33] a note to Huxley reporting on his investigations.[34] Huxley's published paper demonstrates how he was becoming more convinced about transmutation, referring to points in John's research in support of his arguments (Huxley, 1858, p.233-4).

On 18 June, Darwin received a manuscript from Alfred Russel Wallace who was working in the Malayan Archipelago. Wallace was a naturalist and explorer who had maintained a long-distance correspondence with Darwin over a number of years. They had shared elementary thoughts regarding evolution and species, but Darwin had no suspicion that Wallace was quietly working on his own thoughts about transmutation. Wallace finally decided to confide in Darwin during the unusually hot summer of 1858. Wallace's ideas were different – he argued that the environment led to the elimination of the unfit, rather than proposing competition at an individual level. However, they were close enough to panic Darwin. This feeling of despair was not helped by the death of his baby son, Charles, a few days later. Lyell and Hooker stepped in to help their friend. They arranged for a joint publication of Darwin and Wallace's ideas at the Linnean Society on 1 July. Everyone encouraged Darwin to get on with publishing the core of his theory in an 'Abstract' volume that was to become *On the Origin of Species* (Desmond & Moore, 1991, p.468-474).

Darwin frequently corresponded with his national and international network of contacts now to check facts, gather evidence and obtain opinion from those whose judgment he trusted. John was one of those people. In June, Darwin supplied John with queen bee larva and pupae for his dissecting collection, and obtained information from him about the evidence for variation in the number and position of nerves in two species of Coccus John had examined.[35] John prepared a paper on this research for the Royal Society that was read on 9 December 1858 (Lubbock, 1859). Darwin later made reference to John's research in *On the Origin of Species* on six occasions.

Later that summer, Darwin wrote to John reminding him of an observation made by his mother, Lady Lubbock, regarding a variation in certain Pelargonium flowers. He enclosed two flowers and stalks that he asked John to compare. He also asked John to look over his own Geraniums and send him a few trusses that have flowers without marks as well as perfect flowers on the same truss.[36] John, his family and High Elms were all part of Darwin's research laboratory. John also made drawings, with the camera lucida, of worker ant jaws that Darwin had dissected to illustrate an argument in *On the Origin of Species* regarding natural selection (Darwin, 1859, p.241).

Darwin continued to fulfil his mentoring role with John despite his focus being dedicated to *On the Origin of Species*. In November, he commented on John's abstract for his Royal Society paper: 'I congratulate you heartily on producing so profound & philosophical a memoir.'[37] By the beginning of 1859, John's work on insect dissection and anatomical study confirmed Darwin's belief in him as a naturalist of national reputation, as well as a close friend whom he cared about:

'I am grieved to hear of your Brother's accident; but really is most fortunate that it was not worse; what horrid anxiety poor Lady Lubbock must have had & I fear may still have some. Would you some day write me the briefest note to tell me how he goes on, for I shd. really much wish to hear; & I cannot from home, as they are all moving to Hartfield, on account of poor Etty, who is much worse.

'After I last saw you I had a bad attack followed by a second, & I have had to take refuge here, where I shall remain a fortnight & try to get a little strength...P.S. Thinking over your case of ovarium of Pulex, it has occurred to me that you & probably no one but you in England could write a capital paper "on the position of certain anomalous insects in the Nat. System, as judged by their internal organs."'[38]

Darwin condensed his vast work on species and variation into a single volume of 155,000 words. He corresponded and met with John on points of detail. On 8 January, John wrote to thank Darwin for suggesting that he write about the internal anatomy of insects and made a passing comment about how problematic it is to explain how different organs can be in species that are apparently 'very nearly allied', 'unless indeed we cut the Gordian knot, by assuming a creation?'[39] Darwin annotated this letter with clear reference to his work in *On the Origin of Species* and he admits that this difference is a 'very serious difficulty'.[40] John's cautious recourse to 'creation' may also have perhaps raised some doubt in Darwin's mind that John was truly converted to the cause.[41]

On 8 March, Darwin asked John for help by giving him a case of any insect which apparently does not undergo an abrupt physical change after birth or hatching, known as metamorphosis.[42] John replied that 'My own feeling is that all Insects go through metamorphoses but that in some of them a part is passed before birth'.[43] He suggested he visit Darwin on Saturday, an offer which his master welcomed with a request that his apprentice look at a relevant reference to a paper by L. Dufour beforehand so that he might give his opinion when they met.[44] On 21 March, Darwin asked John to give his opinion regarding whether evidence of metamorphosis can be seen in the plates of Huxley's work on young Aphis.[45]

Darwin and John were not in complete agreement in this exchange, but Darwin clearly respected and valued John's opinion as a way of exploring and testing the evidence. He regarded him as an important young convert to what had now become a crusade in Darwin's mind:

'I forget whether I told you that Hooker, who is our best British botanist & perhaps best in World, is a full convert, & is now immediately going to publish his confession of Faith…Huxley is changed & believes in mutation of species; whether a convert to us, I do not quite know. We shall live to see all the younger men converts. My neighbour & excellent naturalist J. Lubbock is enthusiastic convert.'[46]

The focus of that crusade was nearly complete – Lyell had secured a publisher (John Murray) and Darwin had worked on the final proofs over the summer. The title was agreed, the number of copies confirmed at 1,250 and the publication date announced as November. Darwin was about to throw a grenade into the scientific, political and religious establishment and so he retreated to a spa in Yorkshire to escape and observe its impact from a distance. He was not to be disappointed. Hooker, Huxley, Kingsley and Lyell were all supportive, as were the radical atheists such as Robert Grant. Owen, Adam Sedgwick and the Duke of Argyll, amongst others, were not. However, whether people agreed with it or not, they were buying the book. Prior to publication 1,500 copies of *On the Origin of Species* were ordered and a second print run of 3,000 copies was immediately commissioned.

To Darwin's relief, John was also a supporter:

'Have you finished it? If so pray tell me whether you are with me on general issue, or against me. If you are against me, I know well how honourable fair & candid an opponent I shall have, & which is a good deal more than I can say of all my opponents.'[47]

'I am delighted to hear that you are on my or rather our side; I feared from a former note that you were changing your opinion.'[48]

By the time John and Darwin were sitting down to celebrate Christmas with their respective families in the North Kent countryside the stage had been set for a decade of intellectual debate that would help transform Victorian society. The lines of battle were now clearly drawn. *On the Origin of Species* had been published and Huxley and Owen had ignited the debate about man and ape and science and religion. A little known discovery in northern France made in the 1840s was also about to make its mark thanks to a short cross-channel excursion taken by two English gentlemen earlier in 1859.

Prestwich, a wine merchant, and John Evans, a paper manufacturer, both keen geologists and members of the Geological Society, visited St Acheul in April 1859. They examined first hand Jacques Boucher de Crèvecœur de Perthes' claim that he had discovered man-made flint tools of great antiquity. Boucher de Perthes had published his first discovery of a worked flint implement found with the remains of elephant and rhinoceros in the river gravels of the Somme in 1846, but his interpretation of human manufacture had not been widely believed at that time. In Darwin's words:

'My doubt has been… whether the pieces of flint are really tools: their numbers make me doubt; & when I formerly looked at Boucher de Perthes drawings I came to the conclusion that they were angular fragments broken by ice-action.'[49]

Prestwich and Evans agreed that the flint implements had been found in undisturbed deposits and were the result of human manufacture. They reported as such on their return to the London scientific community in papers read at the Royal Society and the Society of Antiquaries in May and June 1859. How could humans have lived alongside elephants in northern France? What connection did this discovery have, if any, to Darwin's ideas published in *On the Origin of Species*? Lyell, and to an extent Darwin, took an interest in the antiquity of man.

John had given a paper about insects on behalf of an absentee contributor at the same Royal Society meeting to which Prestwich presented on 26 May 1859. He would have sat listening intently in the audience, studied the flint implements brought back by Evans and Prestwich and thoughtfully considered the photographs showing these very same artefacts in-situ before they were removed from the ground (Gamble & Kruszynski, 2009). The scientific importance of this groundbreaking discovery would not have been lost to him. A new dimension to the evolutionary debate began to take shape, one that was to draw Darwin's apprentice, John, away from his collecting and dissecting of insects into the world of stone tools and prehistoric archaeology.

Chapter 2

The Flint Dagger

The wooden crates were stacked on the quayside as the grey waters of the mighty Thames swept downriver en route to the North Sea. They were damp, salt stained, scuffed and chipped from rubbing against each other on the lively sea voyage. However, they had survived the journey from Copenhagen across 700 miles of open water and were still in one piece. In the summer of 1863 these crates and their contents lay waiting for a new owner to claim them. All around was hustle and bustle – men shouting orders, chains rattling, ropes splashing into the water, cranes straining, livestock bleating and coopers hammering at casks on the quay.[50]

The creak of the wooden wagon, the skim of its iron rimmed wheels and the thudding clop of the two large brown Shire horses, with their long ruffled manes and white banded noses, came to a stop alongside. The driver dismounted, slips of paper were exchanged and labels checked – 'John Lubbock – Rudolph Puggaard – J. J. Steenstrup'. The crates were loaded on board.

Horse and driver weaved the wagon and its load through the narrow alleys of the London docks, away from the forest of masts on the waterfront with their carnival of sails and coloured flags. Past pubs, sail makers, grocers, provision agents, instrument makers and slop sellers. Aromas of rum, tobacco, coffee, spice, leather and tar filled the air. They navigated their way around dock labourers and watermen, the satin-waistcoated mate, the black sailor with a large fur cap, and the men, women and children holding on to their possessions as they waited to set sail for their new lives overseas (Mayhew, 1861, pp.302-4).

The crates and their contents were destined for a quiet yet fashionable country parish called Chislehurst in North West Kent about ten miles outside of London. The wagon eventually pulled into the drive of a detached red brick house called Lamas that had recently been built in Camden Park. Yet on their arrival these crates were stored to one side. There was no immediate excitement or frenzy of activity to open them and explore the ancient treasures within. The skilfully crafted flint dagger marked with the number '111' rested inside along with its other stone and bone companions, waiting for the moment when their new master would reveal the secrets they held to the wider world.

As these stones and bones were wending their way over land and sea from the capital city of Denmark, their new twenty-nine year old owner and his wife were sitting in a comfortable first class railway carriage travelling in the opposite direction. John and Nelly had left their Chislehurst home and young family behind on 7 July 1863, with bags packed for a two-month Scandinavian tour. They travelled with Mary Arbuthnot, the twenty-three year old daughter of a fellow banker, Sir James Alvey Arbuthnot, who owned the Arbuthnot Latham and Company investment bank.

These were the early years of European train travel. Having crossed the channel by ferry from Dover to Calais they settled in for a train journey via Cologne and Hamburg to Copenhagen:

'I was much astonished to find that there is no railway between Harburg and Hamburg, so that one has to undergo a most disagreeable drive of a couple of hours through a flat, swampy country.'

(Lubbock, E. 1864, p.357)

After another ferry crossing they arrived at Gothenburg on the Swedish mainland. Here they boarded a crowded canal steamer for their four-day excursion through the magical Gotha Canal to the 'Venice of the North', Stockholm.

The Gotha Canal was a feat of early nineteenth century engineering that was thirty years old by the time John and Nelly set off on their journey. It had transformed communication between Western Europe, Scandinavia and Russia by creating a seaway connecting the North Sea to the Baltic Sea. A diverse mix of people and goods travelled along this artery on a daily basis. The party of three would have had plenty of time to admire the majestic Swedish scenery as the steamer puffed its way along canals and across wide, glassy lakes. There would have been secluded valleys, pine forests, stark cliffs and rugged hills, raging waterfalls, quiet villages, old church towers, cornfields and meadows stretched as far as the eye could see. They would also have, no doubt, observed the exotic cauldron of passengers on board:

adults, children and families from Russia, the Baltic states and continental Europe – whether in first class or those packed into second class, forecastle and 'tween decks' with their bags and boxes.[51]

John and Nelly undoubtedly embarked on this trip as a young couple wishing to see more of the world, and to have an adventure in Europe at a time when the excitement of international railway travel was new. They chose Scandinavia as their destination because of John's connection with Darwin and his growing interest in evolutionary theory and the antiquity of man.

Birth of a Prehistorian

By New Year's Day 1860, the second edition of *On the Origin of Species* was already hot off the press and Darwin had sent presentation copies of the first edition to ninety people, including John. The lines of battle were now being drawn as people read and absorbed its contents, deciding on which side of the argument they stood. Darwin was particularly keen to determine his champions within the scientific community,[52] and to identify his enemies! John, Huxley, Hooker, Kingsley, Lyell, Busk and Carpenter were on side. Owen, Sedgwick, Bishop Samuel Wilberforce, the Duke of Argyll and Fitzroy (captain of HMS *Beagle* on Darwin's voyage) were angrily and vocally against, regarding *On the Origin of Species* and natural selection as both ungodly and socially dangerous. Letters and opinions were submitted to publications such as the *Athenaeum* and reviews were written: Huxley in the *Westminster Review*, Owen in the *Edinburgh Review* and Bishop Wilberforce in the *Quarterly Review*. The normally docile BAAS Section D session (Botany and Zoology) held in Oxford on 30 June was packed with over 700 delegates, all sensing something was in the air. The meeting did not disappoint, providing an explosive debate allegedly between Huxley and Wilberforce with Hooker and John active in the supporting cast (Desmond & Moore, 1991, pp.492-9).[53]

At the heart of this war was the question of human antiquity and the proposed relationship between humans and apes. Darwin had for a long time believed in the idea of such an evolutionary relationship with transmutation by natural selection as the mechanism, although he did not mention it specifically in *On the Origin of Species*. Huxley became increasingly explicit about this relationship, using the evidence of brain structure as an important example of continuity between the species. Owen, as we have already heard, refuted this absolutely and, though admitting that man might change, sought a mechanism based upon preordained creation at birth. Owen still believed in the centrality of God and the established Anglican Church in society and his science had to understand nature within that framework.

Huxley sought to brandish Owen a liar blinded by religious contamination and he appealed to the up and coming intellectual thinkers to disconnect science from religion (Desmond & Moore, 1991; Desmond, 1984). They became enemies. When Huxley joined the Zoological Society Council in 1861, Owen left. In 1862, Huxley moved to stop Owen being elected to the Royal Society Council. This heady mix was fuelled further by the publication of *Essays and Reviews* (1860) in which a brave group of progressive liberal theology writers, including the Reverend Baden Powell who was a professor of geometry at Oxford University, explored questions of faith within the new emerging scientific paradigm. However, they were condemned by the established church and two of the authors were found guilty of heresy in 1862, despite attempts by John and other Darwinists to champion their cause (Desmond & Moore, 1991, p.501).[54]

Huxley, Lyell and John were keen to connect humans with the fossil past within the context of this wider debate. Huxley developed an interest in fossil man and became fascinated by the fossilized human remains discovered in the Neander Valley near Dusseldorf in 1856. He also began lecturing on human antiquity and anatomical evolution in Working Men's clubs, introducing a whole new audience to Darwin's ideas. Lyell was nervous about Huxley's belligerence on the ape ancestry question, and remained troubled about the implications of evolution by natural selection for humanity and religion – a concern he had held ever since his friend, Darwin, had shared these ideas with him almost a decade before. Lyell decided to write his own contribution to the human antiquity debate in a book that would bring together the artefact and fossil evidence.

John's interest in human antiquity had been sparked by his visit to Abbeville with Prestwich, Busk and Sir Douglas Galton in April 1860. During this trip, he had met with the now elderly Boucher de Perthes and closely examined the strangely shaped flint implements of human manufacture collected from the gravel-beds of the Somme Valley. The association of these 'prehistoric' implements with the fossilized remains of long extinct mammoth and woolly haired rhino intrigued him. Was it possible to identify a pattern between the type of implement found and the nature of animal and plant remains associated with it?

Although he still published and lectured on biological and zoological matters, John's research activity shifted. He revisited the Somme in April 1862 with Prestwich and Evans before publishing his observations in a newly revitalized Darwinist journal: the *Natural History Review* (Lubbock, 1862a). Seven months before, he had gone to Ireland with Evans to visit the shores of Lough Neagh near Belfast on a hunt for prehistoric stone tools. On a previous occasion, Evans had successfully acquired several, but this time the waters were high and they did not find any flint implements (Evans, 1867). In August 1862, John visited the prehistoric lake village sites in Switzerland with a Swiss archaeologist, Charles Adolphe Morlot, and studied the collection of implements discovered alongside waterlogged animal and plant remains (Lubbock, 1862b).

'I visited nearly all the collections of Lake antiquities & saw five of the Pfahlbauten [lake villages] themselves. Three of them, those at Nernier, Thonon, & Morges in the Lake of Geneva I saw from a boat. The water was from 8 to 12 ft deep, but so clear that I could see quite well, the piles & other things at the bottom.

'We thought we saw a hatchet and I undressed & dived for it. After two or three ineffectual attempts we poked at it with a pole, & it turned out to be only a bit of wood. At Wauwyl…we spent several hours in digging & got three hatchets, three or four implements of bone & a great many bits of pottery & more or less broken bones. There also the beams forming the floor are preserved in the peat, & one could stand, as it were, on the old floor.'[55]

John also began investigating the intriguing large shell mounds, or 'kitchen middens', found in Denmark and Scotland – hence his visit with Nelly to Scandinavia in 1863.

On arriving in Stockholm in July 1863, John and his female companions met up with a select gathering of Scandinavian naturalists, antiquarians and physicians all attending a scientific conference.

'Here has arrived an English antiquarian and natural scientist, Mr. Lubbock, who in a rush wanted to speak to me and Steenstrup. Stp. Left this morning and I am here alone. The Englishman has a young beautiful wife with him. She has deep brown eyes full of jest – you can probably imagine the rest.'[56]

The young couple clearly made an impression on the Scandinavian archaeological and natural history community! They visited museums and took a trip out to Uppsala, the home city and university of the world-renowned botanist and zoologist, Carl Linnaeus.

In particular, they had come to Scandinavia for John to visit two elderly gentlemen. An old correspondent of Darwin, Johannes Japetus Steenstrup, professor of zoology at the University of Copenhagen, and Sven Nilsson, a retired professor of natural history from Lund University who was a zoologist and archaeologist. Now in his seventies, Nilsson had spent many years researching the early inhabitants of Northern Scandinavia and had developed his own theories about cultural evolution. He was one of the first people to bring ethnographic thinking to the discipline of archaeology and proposed several stages of evolution from hunting and fishing communities through to modern society (Nilsson, 1868). Steenstrup had been an important ally for Darwin during the 1850s as he developed his theories of evolution, providing a conduit into relevant Scandinavian research and collections. He was an expert on the Danish kitchen middens (køkkenmødding) and had brought this phenomenon to Darwin's attention in 1852:

'For the last years I have been much engaged with the observations, I have made in company with Professor Forchhammer and Mr Worsaae, inspector antiquitatum danicarum, of the remarquable mounds of shells, which are found on our coasts, and which were formerly regarded as raised beaches, but which we now with certainty can prove to be leavings from the meals of the eldest inhabitants...we found these intermixed with haches, knives and other instruments of flint and bone, fragments of pottery, and many bones of quadrupeds, birds and fishes, which had been eaten... I have already examined more than 30 of these ancient monuments, derivating from the eldest inhabitants and more than 3000 years back.'[57]

Steenstrup had spent years sifting through thousands of mammal, bird and fish bones to reach this conclusion and discovered remains from a number of species now extinct in Denmark, including aurochs and beaver. Within the context of the debate on human antiquity now raging in London during the 1860s, these Danish discoveries took on a new international relevance.

John and Steenstrup had met two years previously when John visited Copenhagen in 1861 accompanied by Busk. Steenstrup had been their guide as John and Busk examined animal bone, human remains and stone tools from køkkenmødding and visited the sites of Havelse and Bilidt. John had returned from this earlier trip in possession of reports about these discoveries written in Danish, with a plethora of ideas in his head and a selection of zoological, bone and stone specimens in his luggage. With the help of a dictionary, he painstakingly translated the reports and published an account of the Danish køkkenmødding in the October 1861 edition of the *Natural History Review* (Lubbock, 1861).[58]

John had also drawn upon and refined the Three Age System of classifying 'prehistoric' artefacts (Stone Age, Bronze Age and Iron Age) invented by C.J. Thomsen in 1819. He proposed a further division within the Stone Age – the Old Stone Age (Palaeolithic) 'when man shared the possession of Europe with the Mammoth, Cave bear, the woolly-haired rhinoceros, and other extinct animals', followed by the New Stone Age (Neolithic) 'a period characterized by beautiful weapons and instruments made of flint and other kinds of stone' (Lubbock, 1865, pp.2-3).

Danish prehistoric archaeology had clearly inspired John on his first visit in 1861 and whetted his appetite. He had been particularly impressed by the collections of Danish stone antiquities he had seen for sale in Copenhagen with a price tag equivalent to between £10 and £20. As early as November 1861 he had asked Steenstrup to look out for such a collection that could be acquired for him and his new collecting partner in crime, Evans (Bangert, 2008).[59] In March 1862, a package of flints and pamphlets had arrived at Lamas from Steenstrup by way of Monsieur Puggaard who acted as a shipping intermediary.

'The flints are much admired by all who have seen them, & your memoir…looks very interesting. I will take care that the copies intended for Busk, Darwin & Owen shall reach their respective destinations. The whole question of antiquities raises much interest here, we are waiting anxiously on the one hand for fresh information from you & from Switzerland, and on the other for Sir C. Lyell's general work on the subject. Our public gradually reconciling itself to the necessary alteration of dates.'[60]

John and Evans wanted more and in April 1862 instructed Puggaard to ask

Steenstrup if he could invest between £20 and £40 for them on further material.[61] This material came in the way of a collection of prehistoric archaeological artefacts found in Denmark and ethnographic material from Greenland which was sold to Steenstrup in early 1863 by a young Danish archaeologist called Wilhelm Boye in order to help pay for his archaeological studies.[62] This collection had been carefully packed into boxes for John and Evans and shipped to England shortly before John and Nelly started on their Scandinavian trip.

The days spent in Stockholm by John, Nelly and Mary Arbuthnot had been hectic and exciting with many new contacts to foster. Towards the end of July 1863, the small party left to travel back to Gothenburg by train from where they embarked on a short excursion into Norway. Picking up carriages in Christiania (now Oslo) they drove cross-country to Drontheim (now Trondheim), and then journeyed by steamship to Bergen and up the Sogne Fjord before returning to Copenhagen via Christiania.

'I do not think there can be a more delightful mode of travelling than in a Norwegian cariole; one has such a lovely sense of freedom; the scenery is so lovely, the air so exquisitely pure and fresh.'

(Lubbock, E. 1864, p.359)

The three companions arrived in Copenhagen on 13 August 1863 to spend a week with Steenstrup, during which time John would have had plenty of opportunity to catch up with news and research at a more leisurely pace. They would probably have spoken about Darwin and his dissatisfaction with Lyell's book on the antiquity of man published in February:

'The Lyells come here this day week, & I shall grumble at his excessive caution: I feel sure that he admits almost fully the modification of species by variation & selection; & yet, though writing at length on subject, is afraid to say so; & he will not serve as guide to anyone. The public may well say, if such a man dare not or will not speak out his mind, how can we who are ignorant, form even a guess on subject.'[63]

About how Huxley more than made up for Lyell's failure to mention ape ancestry by publishing his 'monkey book' as Darwin called it, *Man's Place in Nature*, which presented the case for an evolutionary relationship between man and ape. John would probably also have restated his disappointment to Steenstrup that Lyell 'devotes so little space to Scandinavia and Switzerland that I am thinking of expanding & reprinting my articles'.[64] He would have pumped Steenstrup for information on the latest research and talked about his recent visit to Scotland to examine the kitchen middens near Elgin.

Steenstrup also gave John a small, plain notebook with no title page and no cover, of unremarkable appearance yet of great importance to our story. They would undoubtedly have sat down and talked through the wonderful list of treasures contained within. For this was the catalogue of the collection of stone implements and ethnographic artefacts which Steenstrup had purchased from Boye and arranged the shipment of to John's home through Puggaard two months before.

'Some days ago I sent to our mutual friend Mr. Rudolph Puggaard four boxes with antiquities and ethnographic objects for you. Signed: John Lubbock Esqu. I. II. III. IV and at the same time still three other boxes containing the Antiquities from the Stone= and Bronze=Age for Mr. Evans, signed. John Evans Esq. A. B. C.

'The two collections together are so great and instructive, that I think never before so rich a collection has left Denmark, except the one to Emperor of France as present from our king, and partly consisting of duplicates from the private collections of H.M., partly of duplicates from the Museum of Antiquities.

'Many very good and very seldom objects are in your and Mr. Evans' collections – but what I think is still much better, they are both very instructive.

'A catalogue of your collection written by the former owner, Mr. Boie, is in my hands. Knowing now, that we shall have the pleasure of seeing you soon, I think it must [be] that the catalogue rests here waiting your arrival, and our mutual conversation about its contents.'[65]

The author of the catalogue was a twenty-six year-old 'student' of archaeology with considerable talent but little money. Born in Helsingor, Boye moved to Copenhagen in 1856 to live with a minister when his father died of cholera. In 1857, he contacted Thomsen at the National Museum and asked for some work. Thomsen set him on studying the gold and silver items in the collection. Boye then presented the outcomes of his research to the Danish Society of Antiquaries in 1860. During the early 1860s, he participated in the excavations of important Danish prehistoric sites at Nydam, Thorsbjerg and Vimose.

Boye's collection had been a labour of love for over seven years. Since 1855 he had acquired prehistoric stone, bone and pottery finds from excavations and stray discoveries made across Denmark, including those found by workers when digging drainage ditches. The spidery writing and intimate sketches of artefacts and find locations in the catalogue had been meticulously copied down in English by Boye from his original Danish catalogue, although the original had more information about how Boye acquired the material.[66] Steenstrup would have been able to describe the original find locations in detail for John, and John may indeed have met Boye during his stay although there is no evidence to point to this. The sketches illustrating prehistoric stone tombs in a field on the small Danish island of Moen are likely to have held their attention as they browsed the catalogue (Plate III). Boye collected over 100 artefacts from this site which were now owned by John including human remains, amber beads and manufactured stone implements such as the flint dagger which Boye had labelled '111'. Two years later, John was to dedicate over two pages of his publication, *Pre-historic Times*, to describing the discoveries in this field (Lubbock, 1865, pp.104-7).

John was hooked with Scandinavia, prehistory and collecting stone tools. He and Steenstrup made an excursion to Lund University to see Nilsson's collection of antiquities now housed at the Museum of Antiquities in Lund. It is likely that Nilsson gave John seventeen stone implements during this visit to add to his new 'collection'.[67] Steenstrup and John also visited a small antiquities 'shop' and art gallery at 11 Amagertorv in Copenhagen, owned by a forty-nine year old Jewish art historian, Sally Henriques. Here they bought a dozen stone implements including axes, knives and chisels.[68] Steenstrup donated a few items, and Puggaard also gave him material from a tumulus on Moen and flint tools from Ordrup.[69]

Steenstrup accompanied the party on field excursions to a tumulus at Aarhus and the køkkenmødding at Meilgaard, both on the Jutland Peninsula. It was here that they said their goodbyes and Steenstrup returned to Copenhagen.

'After you were gone we were all quite melancholy & indisposed to do anything, but it was no use being lazy so we went off to the localities which we had visited the evening before. Here we were most successful; you said that "scrapers" were common at Aarhus and I myself found 37 in two hours & a half, of which only 3 or 4 were at all doubtful. In addition we got a lot of short & bad flakes & a great many "knots". Altogether 211 pieces found by me, besides those got by the ladies.'[70]

After all this excitement, the three travellers started for Flensburg in Schleswig Holstein and the return journey back home by train. John would have read and made notes as the steam locomotive passed through the German and French countryside, keen to share his experiences with Darwin, Hooker, Evans and others. Both he and Nelly would have been looking forward to seeing their family again. Perhaps John was also excited by the prospect of opening the treasure trove of artefacts sent by Steenstrup. On his return home to Lamas, he found the boxes there waiting for him.

A crowbar, or similar implement, may have been necessary as a lever to prise open the wooden packaging, but then it would have involved careful and patient unpacking of each item in turn. Some objects would have been more fragile than others and the nagging question in John's mind would be whether they had survived the long sea crossing undamaged. John might have carried out this activity on his own, or it may have been a family affair with Nelly perhaps assisting, for example, in checking off individual objects against Boye's catalogue. The youngest children at least would probably have been instructed to keep a clear distance. John would have admired the workmanship of the beautifully flaked pieces of flint, in awe of objects made by people thousands of years ago. Perhaps he would have studied certain features more closely under a microscope.

As he opened the parcel containing a mottled grey flint dagger marked '111', he may have wondered whether the broken tip was a consequence of the North Sea journey. Turning to the little notebook he had carried back from Copenhagen, he would have discovered with relief that it had been broken when the catalogue entry was written. He would also have read more about the intriguing story behind this object. How it was one of those artefacts discovered on the small island of Moen to the south of Copenhagen, inside a chamber built from large blocks of stone and covered by an earthen mound. It had been buried there thousands of years ago by its makers with three other flint axes, a 'very beautiful lancehead' over eleven inches long, an animal tooth and an earthen vessel. Alongside had been found the unburnt human remains of perhaps its original owner.

John began to write down information about the other items he had collected during the Scandinavian trip, including those from Nilsson,

Puggaard and Henriques, in the same notebook used by Boye. The handwriting in the catalogue changed from Boye's scrawl to John's more rounded style. As he continued the numbering system started by Boye, his prehistoric archaeological and ethnographic collection came into being.

Three boxes would have remained unopened after all this activity, waiting for John to forward them to his collecting companion, Evans, at Nash Mills in Hertfordshire. Before long, they would have been comparing notes about their respective acquisitions. Between September and December 1863, Evans had contributed a further seven arrowheads from Northern Ireland to John's newborn collection.[71] This growing friendship with Evans was of particular importance to the birth of prehistoric archaeology and the wider history of science. They shared a mutual interest in flint implements, and their family backgrounds and personal situations were also similar in many ways. Both were members of business families that formed part of an emerging entrepreneurial elite, although, it was clear that John's family wealth and connections were more established than those of Evans, who gained his position through a combination of hard work and advantageous marriage. Both Evans and John worked hard in their respective family businesses, regarding their scientific work as a leisure rather than professional activity. Both were married with children and had begun work at an early age: neither of them had a university education.

During the early 1860s, John was creating new friendships with Evans, other antiquarians and scientists that would last a lifetime. These friendships would champion Darwin's ideas and help change the way we look at the world. His collecting activities and interest in prehistoric archaeology played a central role in developing these relationships.

Chapter 3

The Bronze Bracelets

T he two friends casually walked back down the trail admiring the beautiful alpine mountains and lake scenery surrounding them. Tired and dirty after a long day, they talked excitedly about the contents of a box they carried with them. Wrapped up inside were two delicate bronze bracelets and other ancient artefacts from a 2,500 year old Iron Age grave excavated by their own hands a few hours earlier. As they descended that evening back to the Austrian village where they were staying, it is likely that topics of conversation included looking forward to a long drink and a hearty dinner. The two companions might also have talked about how they could acquire more of these wonderful treasures for the British Museum and their own growing private collections at home in England.

John and Evans had boarded a train in London a few days before, accompanied by John's wife, Nelly, and travelled by way of Paris to arrive in Salzburg during April 1866. Here, in this romantic city steeped in history, Evans at least had taken time to visit the museum and study the ornate prehistoric metalwork discovered in nearby mines.[72] The three travellers then left Salzburg in a horse and carriage on a journey into the Alps to the place where this metalwork had been found, the picture postcard village of Hallstatt.

Hallstatt is an alpine gem now classified as a World Heritage Site by UNESCO. In 1866, the party would have crossed the fjord-like Halstättersee on a tiny ferry to discover a small settlement of timber houses built on a mountainside ledge and set on wooden piles anchored into the lake floor. No road led in or out, and the village was connected to the outside world by water and a few, very narrow, mountain trails. Yet, connected to the outside world it most certainly was. For many thousands of years people had lived in this apparently inhospitable yet beautiful place because of a mineral hidden deep inside the mountains behind. Sodium chloride (NaCl) was the commodity that the salt miners of Hallstatt exported to the wider western world. For over 500 years it has been transported as brine through a pipeline that stretches forty kilometres to join the river networks of Europe at Ebensee.

It was this salt mining activity that also brought John and Evans to Hallstatt. For they were interested in the prehistoric miners who had exploited the enormous reserves of salt over 3,000 years ago and whose remains had been so well preserved in the unique setting of this remote mountain village. On 19 April, they took their first hour-long hike up the mountain trail to the prehistoric cemetery site with 'a commanding view' that gave its name to an Iron Age 'Hallstatt' culture that is still used in archaeology today. The next day, the two comrades returned to the site and dug up the remains of a burnt body, finding bronze bracelets, brooches, an iron celt with part of its handle and evidence of fine woven material on it, bronze rings and portions of what looked to be spearheads.[73]

After the excitement of these discoveries, the party left Hallstatt and travelled together as far as Vienna where they went their separate ways. Evans headed back to London via Germany, and Nelly and John travelled on to visit the Italian cities of Venice, Milan, Florence, Rome and Naples. When they eventually arrived home and unpacked their cases, the bracelets and other finds would have been taken into John's place of study where they were carefully numbered and their circumstances of discovery recorded. With some pride, he entered the words 'Found by me' and 'Found by us' in his catalogue.[74]

John and Evans' interest in Hallstatt had been piqued by their short visit, and they were especially struck by its potential as a source of prehistoric artefacts for their collections. Johann Georg Ramsauer had excavated over 1,000 graves since he was first commissioned to do so by the Austrian government in 1846, but these excavations had stopped in 1863. From their short visit, John and Evans could see there was still a lot to discover and acquire for collections back in England.

By 1866 John had serious wealth to play with. One year previously, his father had died and John, as the eldest son, inherited the family estate and business. Despite the banking crisis of May 1866, which eventually required John and Nelly to cut short their European trip, they felt able to invest with Evans in restarting excavations at Hallstatt. Arrangements were made with the local government official, Joseph Stapf, for work to recommence at their expense and for any discovered finds to be sent to them. Within a month of their visit, Evans and John had parted company with forty Austrian florin banknotes and Stapf had already opened two graves. Stapf wrote to let John and Evans know that 'several beautiful and interesting things' had been discovered which were to be sent later on.[75]

Work was carried out under this arrangement during the summers of 1866-8, but Stapf was not an easy business partner. He haggled about the money required to pay for the work and items found.[76] The Austrian Crown Prince Rudolph paid a royal visit to the cemetery in October 1866 and was shown the excavated finds.[77] As a result, Stapf had to present him with many of the best pieces, much to the disappointment of John and Evans. However, they still received a considerable amount of material. The first consignment, sent on 9 January 1867, consisted of 82 listed iron, bronze and amber artefacts, several pieces of bone and several blue glass beads.[78] John and Evans shared out the collection and donated some material to the British Museum. A further two consignments from Stapf followed in February 1868 and January 1869.

In total this collaborative project cost them 750 Austrian florin, but its true value was far greater than the quality of collections and evidence acquired. The memories of that day spent excavating at Hallstatt would have stayed with both John and Evans for the rest of their lives. This shared experience, and the many others represented by artefacts in their collections, was an integral part of their unique and significant friendship.

Significant Friendships

Friendship and brotherhood were at the heart of the Darwinist movement in the late nineteenth century. A group of men with similar values and beliefs, and ably supported by the women in their lives, fought side by side in the battle for evolution. They did not always agree and some friendships were closer than others, but they held shared experiences and were connected through their ambition for change. For John, many of these personal relationships were forged during the early years of the 1860s, with his prehistoric archaeology interests and collecting activities a recurring theme in these associations.

THE BRONZE BRACELETS

As well as collaborating in Scandinavia and Austria, John shared a fascination with Evans for the 'bone caves' of southern France. Remains of reindeer and other Arctic Tundra species had been discovered in association with various flint implements of human manufacture in caves and rock shelters nestling in the limestone cliffs of the Dordogne region. These intriguing archaeological sites were being excavated by Henry Christy and Edouard Lartet, and so at Easter 1864 John and Evans set off on a trip by ferry and train to see this work first-hand. Their main destination was a small village nestled in the Vézère Valley called Les Eyzies-de-Tayac. As they rounded a bend in the road and crossed the bridge over the river into the main settlement they would have been struck by the dramatic river cliffs lining the side of the valley, sculpted by wind and rain over thousands of years, and by the village houses and shops built into these massive monuments to natural erosion.

'Apart from scientific interest, it was impossible not to enjoy the beauty of the scene which passed before our eyes as we dropped down the Vézère. As the river visited sometimes one side of its valley, sometimes the other, so we had at one moment rich meadowlands on each side or found ourselves close to the perpendicular and almost overhanging cliff. Here and there we came upon some picturesque old castle, and though the trees were not in full leaf, the rocks were in many places green with box and ivy and evergreen oak, which harmonized well with the rich yellow brown of the stone itself.' (Lubbock, 1913, p.319)

They spent the next few days visiting the now famous prehistoric cave sites of Laugerie, Moustier, Les Eyzies, Badegoule, La Madeleine and Bourdeilles. John collected 'rude flint implements' of various shapes and sizes, including cores, scrapers, flakes and lanceheads. The Vicomte de Lastic also presented him with a hammer from Bruniquel. When he returned home with his boxes of stones, one of his first tasks would have been to record the details of these new acquisitions in his collection catalogue and to assign the artefacts with their respective catalogue numbers so he would not forget where they came from.[79]

In September of the same year, John and Nelly attended the BAAS meeting in Bath. During this visit Evans, Francis Galton (Darwin's cousin) and John took a short walk to Little Salisbury Hill nearby and picked up prehistoric arrowheads and flint flakes lying on the ground. Both Galton and

Evans donated the items they found to John's growing collection,[80] while Galton also donated some ethnographic material from his travels in Africa.[81] Although none of the 'scientific' supporters of Darwin – Tyndall, Huxley, Hooker and Busk – held their own collections of prehistoric artefacts, the new activities of their friend, John, kept them connected to this developing line of research. And John was most definitely developing strong relations and friendships with these individuals. Huxley and Hooker had both been impressed with John's performance at the BAAS in June 1860. Darwin expressed his pleasure at their comments in a reply to Huxley, saying 'I am glad to hear about J. Lubbock, whom I hope to see soon, & shall tell him what you have said.'[82]

In March and April 1862, John visited Darwin at Down at least twice, once with Hooker and the next time with Busk and Huxley (Burkhardt *et al* 1997, p.666). Once Nelly and John were ensconced in their own home, having moved out of High Elms in August 1861, they had the freedom to organize their own social gatherings and parties. In February 1862, for example, they invited Darwin to come over and meet the Busks and Hooker.[83] In March John visited Hooker at the Royal Botanical Gardens, Kew, creating a great impression. Hooker wrote to Darwin of the visit 'I am as much charmed as you are with him. We hope he will bring Mrs. L. to Kew. I like her so much.'[84] To which Darwin replied 'I am pleased to hear that you like Lubbock and Mrs. L.; he is a real good fellow & she is a charmer.'[85]

In July 1862, John was persuaded by Huxley and Tyndall to spend ten days with them climbing in the Swiss Alps. This was his first trip to Switzerland and it was to be a memorable one for many reasons. John's principal interest in visiting the country was to investigate first-hand the ancient Swiss lake villages; preserved in waterlogged conditions they contained stone implements, pottery and animal/plant remains of great interest. He planned to meet up with the Swiss archaeologist, Morlot, to examine these sites in late July. However, first he had bravely agreed to meet with Tyndall and Huxley at a new hotel owned by Monsieur Seiler and located at the foot of the Rhône glacier. He waited for their arrival and when they did not show, he gave up and started into the mountains with a guide bound for Brientz. Tyndall and Huxley had been delayed by bad weather and had been forced to take a different route than originally planned through the mountains from Grindelwald. Tyndall described how as they 'rounded an angle of the Mayenwand, 2 travellers suddenly appeared in front of us' – being John and his local guide (Tyndall, 1883, pp.141-152).

They all returned to the warmth and comfort of Seiler's hotel, with its distinctive wrought ironwork and striking red painted window shutters, and trained for three days on the Rhône glacier. On one of these days, John and Tyndall decided to attempt the Galenstock. At 11, 755 feet above sea level and with ice and snow at its peak, the Galenstock is the guardian of the Rhône glacier and was first climbed in 1845.

> 'The sky was clear and the air pleasant as we ascended...the higher snow-fields of the great glaciers are altogether beautiful...We reached the top, and found there a gloom which might be felt...But suddenly, in the air above us, the darkness would melt away, and the deep blue heaven would reveal itself spanning the dazzling snows.'
>
> (Tyndall, 1883, pp.142-3)

Having passed his first mountain climbing test, John was ready to attempt the Jungfrau with Tyndall and his guide. They moved on with Huxley to the Hotel Jungfrau on the Eggischhorn and prepared for their ascent. However, on this occasion they did not make it to the summit. Traversing carefully across the smooth blue-white surface of ice, John would have been captivated by the sheer dramatic power of this cold and dangerous glacial world. He would also have begun to understand how in mountaineering trust in your companions is paramount. After a few hours walk on the glacier, the party of three noticed a solitary human being standing on the lateral moraine in the distance. He was one of the two porters sent on ahead to carry their provisions for an overnight stop. The other had fallen into a 'wide and jagged cleft' in the ice after a snow bridge he was walking across collapsed under his weight. The travelling companions spent the rest of their excursion rescuing the unfortunate porter from his predicament (Plate V). It was an exhausting and emotional experience – one that John and Tyndall would remember for the rest of their lives. These are the moments shared that create unspoken bonds in a friendship:

> 'I do not think I have ever seen anything so good of you. There is sweetness in the face and earnestness in the eye, and there is ease in the position. I will keep this and [like] instead of the original.'

Tyndall prefers this reverence and earnestness to the

> 'brilliance which dazzles and delights so many of your Friends. But the union of both – this substance of earnest[ness] and the [rippling]

surface phenomena of brilliancy – is I have little doubt destined to bamboozle the world some day.'[86]

On the Saturday before Christmas in 1863, John met with Busk and Tyndall for dinner at the Athenaeum Club on Pall Mall, after which they caught the train to Swindon where they stayed overnight. Over the next three days the trio walked along the Marlborough Downs to Devizes and Melksham. They caught the train back to London and dined at the Athenaeum again. During this excursion, John collected a scraper and arrowhead he found on the ground. Once he got home he added them to his catalogue marked with the numbers '253' and '254'. At Christmas 1864, John went for a short walk in Derbyshire with his alpine travelling companions, Tyndall and Huxley.[87]

Tyndall, Darwin, Huxley, Hooker and others actively encouraged John to present his theories and observations regarding the antiquity of man in the wider public domain:

> 'My dear Mrs. Lubbock – no doubt you have already been informed of the success of last night's lecture; but probably your husband will not tell how successful it was…I could number on my finger-ends the lectures which have been equally successful during ten years of my connection with the Royal Institution. At the conclusion there was a loud and long-continued outburst of applause…As for me, who, it must not be forgotten, stirred up his mind to the performance, I feel quite bright by the reflected radiance.'
> (Letter from John Tyndall to Ellen Lubbock, 28 February 1863, reproduced in Hutchinson, 1914, pp.58-9)

Behind the scenes sat John's mentor, Darwin, observing this activity from his home in Kent and deliberately keeping a distance from the public human antiquity debate. Occupying himself with further research into the domestication of species and orchids, Darwin still made time to encourage and support his apprentice in this new area of academic interest. He spoke with John about the discoveries in northern France and trusted in his first hand experience:

> 'John Lubbock tells me that the flint tools in France are found in such vast numbers, in Peat, that M. Boucher de Perthes told him that he might take as many as he liked. These facts, to my mind, remove one of the greatest difficulties of the case of the gravel-beds – celts,

namely their surprising numbers. I do hope that you will go to France again, & give us lots of Sections. I found that until J. Lubbock drew me a rough section I did not in the least understand their position; & hardly anything seems known about the extension of the beds of gravel, clay & c. or their manner of formation. The case seems to me to deserve not day's but month's of work.'[88]

Darwin may have shared information that he had received from Colonel Erskine in Aberdeenshire, who had found vast numbers of arrowheads at one place on his property.[89] He was disappointed to miss a paper presented by Edouard Lartet and Leonard Horner on 16 May 1860 at the Geological Society about the evidence for humans having existed alongside animals that had since become extinct,[90] and would undoubtedly have picked John's brains about it the next time they met. John still undertook some natural history research with Darwin commenting on his articles and using material supplied by John in further editions of *On the Origin of Species*.

However, when John and Nelly moved to Chislehurst in the summer of 1861, it became less easy to see Darwin. Although, at least Chislehurst was not as far away as Brighton, where they nearly moved! The move brought Darwin to express his feelings on their friendship, writing to John, 'your talking of living in Brighton made me realise how much I have enjoyed your friendship & what a loss your absence would be to me'.[91]

They both missed their regular conversations and the ready access to each other's advice. In January 1862, John was desperate to speak with Darwin to get his opinion on a controversial statement he proposed to include in a lecture on the antiquity of man. The paper made direct connection to Darwin's work on species.

'John doesn't know I am writing this, but I have heard him lately so often wish he 'could only have a talk with Mr Darwin.'[92]

Unfortunately, on this occasion Darwin was too ill to reply in time. Although the pattern of their friendship changed during this period (more letters and less frequent visits), the bond between master and apprentice remained strong. At Easter 1864, Darwin was one of the first people in England to donate a prehistoric implement to his apprentice's new collection.[93] In the autumn of that year he also donated the unique bone harpoon from Tierra del Fuego that he had perhaps collected all those years ago on the voyage of HMS *Beagle* from Jemmy Button.[94]

By late 1864, momentum was building around John's collection and a whole new world of study, networks and connections was opening up to him and his Darwinist companions. Joseph Anderson in Scotland, Henry Christy, Dr Gordon, Richard Entwisle (a cousin from a Manchester banking family) and Mr Lukis all donated prehistoric flint implements in late 1864.[95] Dr John Rae, a famous Arctic explorer of the mid-nineteenth century, donated an 'Esquimaux' cheek-stud from the Mackenzie River in northern Canada.[96] John also started to buy ethnographic implements from a London-based dealer also used by Evans, William Wareham. These early purchases included a fishhook from the South Pacific 'supposed to have been brought back by Captain Cook', an Australian boomerang and an 'Esquimaux' fishing line made from whalebone.[97]

Now was the time when Darwin's 'brothers in arms' would change the world of science and society. In 1861, a group of them had already started to seek control of the debate by buying the *Natural History Review* and converting it into a voice for Darwin. John, Huxley, Busk, Carpenter and others became editors of this series over the next three years. In late 1864, John, Evans, Huxley, Spencer, Tyndall, Galton, Busk and others bought one share each in the *Reader* to replace the *Natural History Review*. At the same time, nine men met at St Georges Hotel in London to set up a new and elite dining club with the purpose of championing Darwinist agendas within the scientific establishment. John, Huxley, Hooker, Busk, Tyndall, Edward Frankland, Thomas Hirst, Spencer and William Spottiswoode became the founding and only members of the 'X Club' – a very exclusive group destined to take on and transform London's scientific elite.

Chapter 4

The Stone Axe Fragments

Approximately 5,000 miles south west of London, William H. Campbell sat at his desk in Demerara, British Guiana, to compose a letter to his English contact, Hooker (Plate V). Feeling the welcome relief of the sea breeze on his face and ignoring the early evening rain forest chatter of birds and monkeys in the distance, he picked up his pen and wrote the date '9 December 1867' on a small sheet of note paper. Campbell was a regular correspondent with Hooker, and a supplier of plant and seed specimens for the Royal Botanical Gardens, Kew. On this occasion, he informed Hooker that he would soon be receiving a shipment of stone axe fragments and pottery which Hooker might want to send on to John:

> 'You have asked me to get you some stone implements if I should come across them. Curiously enough, I lately showed an Indian from the Rupumani[98] River what sort of things I wanted, and he said that in about six months he would return and bring me what he could find – a week or two ago he came back with about 100 fragments of stone axes & c. which I intended to send to Sir John Lubbock, to whom our lamented friend Christy introduced me – I shall now send them to you, and if they are of any value I have no doubt you will share them with Sir John Lubbock if he desires to have any.'[99]

These stone tools had already travelled a long distance from the South American interior to arrive at the coast. Stashed in the bowels of a dug out canoe they had sat alongside other goods for barter, sale and exchange. Skilfully transported by the Macuxi Indian pilot through the white waters and shallows of the Rupununi River, they had passed under the shadow of the Kanaku Mountains of southern Guiana and continued into the expansive Essequibo River that slowly flowed towards the sea. Walls of green impenetrable rain forest had stood guard on either side as they travelled downstream, interrupted occasionally by a broad flat Savannah in which horned cattle grazed. As they approached the coast the landscape transformed into a uniform pattern created by the 500 acre plantations growing sugar cane, coffee and cotton bushes; established by the Dutch West India Company and now part of Britain's global empire. The thirty degree Celsius temperatures, sixty-five percent humidity and torrential late afternoon downpours through which the indentured Indian plantation workers toiled did not halt the journey of our stone axe fragments. The anonymous Macuxi Indian, who had originally sourced this collection of material soon to arrive in England, was oblivious to the climate, the flies in his mouth and up his nose, and the constant buzzing of 'museketoes'. As he neared his final destination, he navigated through the busy riverboat traffic that carried coffee, sugar, cotton, rice, plantains, cassava, timber and rum to their respective markets (Bell and Balbis, 1832).

Once the stones were in Campbell's hands, they set sail on the second and longest part of their journey, accompanied by two 'money pots' or seed capsules containing seed from Captain Kerr:

'I write merely to mention that I have sent you by the R.M. Steamer leaving tonight, two boxes containing the stone axes and other articles of Indian origin mentioned in my last letter. They are addressed to the 'Admiralty' as usual.'[100]

Campbell had also packed two fragments of axes fitted into rough wooden handles for illustrative purposes, in order to show modes of fastening and use.

Hooker presented these artefacts to his friend, John, in February 1868. This gift is the first entry recorded in the bound notebook with marbled cover that became the second volume of John's 'Catalogue of my Collection':

'580 Collection of stone implements & pottery from the Rupununi River Br. Guiana forwarded by W.H. Campbell & Pres. By Dr. Hooker Feb. 1868.'[101]

Brothers in Arms

Nearly four years before, on the evening of 3 November 1864, John and Hooker had both walked through the doors of the elegant St George's Hotel on Albemarle Street, in the heart of Mayfair and close to the Royal Institution. Inside they took off their coats and hats, and sat down for dinner with six companions: Huxley, Busk, Tyndall, Hirst, Spencer and Frankland. However, this was no casual occasion between friends. It was the inaugural meeting of a new dining club that was to transform the intellectual establishment, one that brought together a select group of people who supported Darwin, his theories of evolution and who held the belief that science should be separated from religious dogma. Huxley had long desired to set up this 'club' during the early 1860s to strengthen the bonds of friendship within the Darwinist community. He finally succeeded in getting all of his brothers in arms together around the same dining table. From that point onwards during the 1860s they met regularly at St George's to swop news, debate ideas, eat and drink well and plot ways to change the scientific world (Barton, 1976). Ironically, given the intellectual heavy weights sitting around that table, they struggled to find a title for their club. One legend has it that the 'X Club' was proposed casually by one of their wives in response to the fact that it had no name. Hutchinson suggested that the name was the mathematical symbol for an unknown quantity, and the only rule was to have none (Hutchinson, 1914, p.63).

Membership of the Club was exclusive. The demanding criteria for membership, as recorded on one occasion by Spencer, included an exceptional mental capacity and an existing relationship on terms of intimacy with other members. Spottiswoode, a mathematician and partner in the King's Printers firm of Eyre & Spottiswoode, was the only person not present at the first meeting who was also invited to join. Carefully selected guests were frequently invited to individual meetings. The sense of closeness between members was also reflected in the social dimension of the X Club. Between 1865 and 1875, members and their wives went on a ritual social outing every June along the Thames Valley. On 23 June 1872, for example,

John went to Windsor with Nelly, Huxley, Tyndall, Mr and Mrs Spottiswoode, Mr and Mrs Hooker, and Spencer.[102]

Members were also carefully selected to ensure that the Club had access to and influence in key areas of society: the City, Parliament, medicine, industry, the liberal church, the University of London and the scientific establishment. Though the majority of members were professional academics, it also included John, Spottiswoode and Spencer. John in particular provided a point of access into the new wealthy classes of Victorian England who had made their money from the Industrial Revolution and financing British imperial expansion. The last formal meeting of the X Club took place in 1893 and was attended by Hooker, Frankland and John. Between 1864 and 1893 it met 240 times, with meetings more frequent in the earlier years. With the passing of time, it became increasingly difficult for members to meet due to the pressures of other commitments, ill health and occasionally rivalry in the later years.[103]

However, that inevitable decline was all in the future; the 1860s and early 1870s were the time of the X Club. Their proceedings were held in private, and the membership of nine soon gained a reputation within the wider establishment of being a powerful yet secretive scientific fellowship (Barton, 1976, pp.1-12; Desmond, 1994, pp.327-35; Jensen, 1991, pp.143-65). It was no mere talking shop, and the group was intent on converting their views into action. From the outset, they began to use their growing contacts and influence to achieve their aims.

We will explore in a minute how the X Club took on the challenge of transforming the power base in London's scientific community. However, before doing so, we need to bring another key player into the picture, Evans, and introduce the archaeological entourage that clustered around John and Evans at this exciting time in the history of ideas.

Evans was not a member of the X Club and it is unclear whether he was invited as a guest to any of their meetings. However, he was a friend of John and shared a common interest in exploring how prehistoric archaeology could shed new light on human antiquity and theories of evolution. Evans had made his name in scientific circles, alongside Prestwich, in 1859 when he publicized the discoveries of ancient flint implements at St Acheul in northern France. He undoubtedly met and conversed with Huxley, Hooker, Busk and Tyndall (and perhaps Hirst, Spencer and Spottiswoode) at meetings of the Royal Institution, the Geological Society and the BAAS. He was a partner in the procurement of the *Reader* in 1864, with John, Huxley, Spencer, Tyndall and Spottiswoode. This purchase was an attempt by the fledgling X Club to champion a liberal approach to science, religion and social affairs. Although it only survived for eight months, it is significant that Evans was asked and felt able to participate in a venture so closely associated with the X Club and its mission.

John and Evans were also pivotal members of an informal grouping eager to encourage change and reform within the traditional antiquarian establishment. Although never formally convened in the same way as the X Club, the dynamic 'Lubbock-Evans network' worked to change the Society of Antiquaries and the Archaeological Institute from 'private gentlemen's clubs' into professional scientific societies (Bowden, 1991, p.46; Chapman, 1989; MacGregor, 2008; Morris, 1996). Augustus Lane Fox (later Pitt Rivers) was a military officer in the army who had fought in the Crimea and developed an interest in archaeology and ethnography whilst posted overseas. Franks was an obsessive collector of medieval artefacts, an assistant in the Antiquities Department at the British Museum, and director of the Society of Antiquaries in London. William Boyd Dawkins was a geologist and archaeologist who had first started collecting fossils at the age of five from the local colliery spoil heaps near his home in Welshport. These three worked with John and Evans to help engineer change. Pitt Rivers, Evans, John and Franks supported each other's candidacies for office in the Society of Antiquaries and the Archaeological Institute (Chapman, 1989). Evans and Franks were long-standing Fellows of the Antiquaries, both elected during the early 1850s. They facilitated the election of John and Pitt Rivers in 1864, and by 1867 all four of them sat together on the Society's council.

The X Club and the Lubbock-Evans network shared many things: their desire to place science and research on a professional footing, their ambition to transform and open up the scientific and intellectual establishment in London, and their active interest in human antiquity and the analysis of surviving fossils and artefacts. These two groups of friends were physically connected by the presence of a single man, John.

During the two decades following Darwin's seminal publication of *On the Origin of Species*, the agents of change in archaeology and science worked together to lobby for a new intellectual world order.

The X Club was particularly interested in transforming the powerful heavyweight of the London scientific community, the Royal Society. Founded in the mid-seventeenth century as a radical society by renowned intellectuals including Christopher Wren and Robert Boyle, and with the patronage of King Charles II, the Society set itself the objective of 'improving natural knowledge'. Approximately 200 years later, it had become, in the eyes of Darwin and his disciples, a dangerously conservative organization under the leadership of its elderly president, Edward Sabine. By the end of the nineteenth century, the Royal Society had been transformed into a forward thinking, professional and reenergized scientific body.

The X Club members and Evans were vital agents in causing this change, not only through their publications and lectures but also in their work to infiltrate and influence the establishment. Darwin had been elected as a Fellow of the Royal Society on 24 January 1839, Hooker was elected almost ten years later in 1847, Busk and Huxley in 1850, Tyndall in 1852 and John in 1858. Evans was elected a Fellow in 1864, undoubtedly with the support of John and his X Club companions. By the 1860s, John, Busk, Huxley, Tyndall and Hooker were all active members of the influential Philosophical Club of the Royal Society, a dining club established in 1847 with clearly stated aims 'to check any retrograde tendencies in the council of the Royal Society, to stimulate the intellectual activity of its members, and to strengthen the influence of science in Britain.' (Barton, 1976, pp.10-11).

In November 1864, the X Club succeeded in persuading the Royal Society to award the Copley Medal to Darwin, against the wishes of Sabine (Desmond, 1994, pp.327-35). First awarded in 1731, this medal is still given annually for outstanding achievements in research in any branch of science. Tyndall was also awarded the Rumford Medal in that year. During the period 1860-68, at least two members of the X Club plus Evans were elected each year to serve on the Royal Society council. Their successful nominations seemed to take place in planned rotation:

1860-1	Huxley, Spottiswoode, Tyndall
1861-2	Lubbock, Spottiswoode, Tyndall
1862-3	Lubbock, Hooker
1863-4	Hooker, Busk
1864-5	Busk, Hirst
1865-6	Hirst, Frankland, Spottiswoode
1866-7	Spottiswoode, Huxley
1867-8	Huxley, Evans
1868-9	No X Club members on council
1869-70	Tyndall
1870-1	Tyndall, Lubbock, Hooker
	Spottiswoode – Treasurer and Vice President

Their interest and influence in the Royal Society appears to have waned a little towards the end of the 1860s, but in October 1870 there is evidence for a sudden burst of X Club activity. The long-standing treasurer of the Royal Society died, creating a vacancy which Darwin's brothers in arms grasped as an opportunity for change. They engineered Spottiswoode's nomination as Treasurer and Vice-President. In the same year, three members of the X Club – John, Hooker and Tyndall – were elected to the council. John, Hirst and Busk were elected to the council in 1871, and Busk, Hirst and Hooker in 1872; Huxley became Secretary and John a Vice President. In 1873, X Club dominance of the Royal Society was secured when Hooker was elected as President, Huxley as Secretary and Spottiswoode as Treasurer. Evans was elected to the council. Hooker served a five-year term as president and was replaced in succession by Spottiswoode in 1878 and Huxley in 1883.[104]

The Ethnological Society became another target for the X Club and Lubbock-Evans network during the 1860s. It had been founded in 1843 as a scientific society primarily for military officers, civil servants and members of the clergy interested in studying the culture and biology of non-western societies. By 1859, however, the whole purpose of ethnology was unclear and the role of the Society was in crisis with only sixty members on the books (Chapman, 1989). At the same time, Darwin's supporters were looking for a public forum within which they could encourage debate and the sharing of knowledge about their new scientific approach to human evolution. Before 1863, only Spottiswoode, Evans and Pitt Rivers were members of the Society. By 1863, John, Franks, Huxley and Busk had signed up and John was elected President. Busk, Huxley and Spottiswoode kept him company on the council, and in 1864 Evans joined them.

Clearly a concerted move had been made to take control of the Ethnological Society. John worked closely with Evans, Pitt Rivers, Franks, Busk, Huxley and Spottiswoode to achieve this. Huxley wrote to John in May 1863:

'I am very pleased to hear from Busk that you are to be the new President of the Ethnological Society. Of course under these circumstances I shall become a member and do my best to help you though, as you know, that best is likely to be little enough. Let Rolleston and...all the good men and true know of your intention.'[105]

The Society rapidly gained 200 new members, who were drawn to this exciting new mission, while the Ethnological Society and its published Transactions (which John helped to edit) continued to act as an important Darwinist public mouthpiece through the 1860s. The X Club and the Lubbock-Evans network used the Ethnological Society to develop and promote a scientific and professional approach to studying human evolution, and to actively oppose their critics.

Owen, Bishop Wilberforce and the Duke of Argyll were not the only opponents with whom they had to engage.[106] A powerful group of radical scientists led by the charismatic James Hunt and Richard Burton also attacked Darwin's ideas and objected strongly to events at the Ethnological Society in 1863. Hunt and his followers then left the Ethnological Society and set up the Anthropological Society of London. In particular, they objected to the idea that all humans had evolved from ape ancestry and represented a single species. Instead, they argued for 'polygenesis' – that human races living across the globe in the late nineteenth century were representative of several human species. They regarded non-western peoples as culturally and biologically inferior, incapable of intellectual development, and towards the latter part of the 1860s became known for their extreme racism. The Darwinists also painted a depressing and patronizing portrait of non-western peoples, reflected in their use of the labels 'primitive' and 'savage' to describe them. However, their belief in 'monogenesis' meant they regarded non-western societies as open to change and capable of adopting Western culture and education if given the opportunities to do so (Bowler, 1989, pp.228-37; Stocking, 1987, pp.228-51).

During the mid-1860s the philosophies of both Societies were distinctive enough to justify their parallel existence. However, by the late 1860s, for practical and political reasons, the two Societies agreed to merge into a single organization – the Anthropological Institute of Great Britain and Ireland. Yet again, the X Club and Lubbock-Evans network ensured that a core group of Darwinists took control. John, as the candidate most acceptable to both societies, was elected the Institute's first president in 1871. The first officers and council consisted of a fair representation from both the Ethnological and Anthropological Societies. However, the elections for the presidency and council in 1873-4 created a 'standoff' between the two groups which the Ethnological Society supporters appear to have won,[107] and for the remainder of the decade they dominated the Institute.[108]

During the 1860s the BAAS became another arena in which the

Darwinists had an ever-growing presence which was exploited for their own ends. Every year the Association's meeting was held in different cities across Britain; this provided an opportunity to broaden peoples' geographic horizons and encourage networking between like-minded individuals outside of London. Before 1866, the contributions on human evolution and antiquity were split across three different Sections, or programmes of activity: Section C (Geology) where the reports on flint implements and ancient human remains would be tabled, Section D (Biology), and Section E (Geography and Ethnology). In 1864, the Darwinist brethren had sought to rename Section E as 'Geography, Ethnology and Anthropology', but failed to convince the scientific bastion of its intellectual significance. The following year, they failed in their attempt to create a completely new Section just for Anthropology.

In 1866, they decided on another tack. Huxley was elected President of Section D and used this opportunity to drive through the creation of a Department of Anthropology as part of this Section. The first President of this Department was Alfred Russel Wallace. John and Nelly travelled up to Nottingham with the Spottiswoodes, Hirst, the Busks, Franks, and a good friend of Hooker, Archibald Hamilton. They would no doubt have given active support to this initiative that sought to put the subject of human cultural and biological evolution firmly on the establishment map.

The Department of Anthropology took a while to gain momentum in subsequent years. In Dundee in 1867, for example, the Department programme was cancelled because of fears that its liberal content would offend the 'conservative Scottish establishment'. However, in 1871, the Department of Anthropology was re-established under Section D and remained in existence until 1883, after which a new Section, H, was created specifically for Anthropology. The Department hosted twenty-five to thirty papers annually that focused on human cultural and physical evolution: ethnography, prehistoric archaeology, geology, biology and language studies. The Presidents during those years included Evans, Pitt-Rivers and Huxley (Sillitoe, 2005). It is no coincidence that 'Anthropology' reasserted itself at the BAAS in 1871, in the same year that the Anthropological Institute of Great Britain and Ireland was created under John's presidency.

The emerging intellectual Darwinist powerhouse had an international dimension and played a key role in setting up the International Congress of Anthropology and Archaeology. The first of these annual meetings was held in 1866 and chaired by the Swiss geologist, Pierre Jean Edouard Desor. In 1868, the Congress met in Britain. The membership of the organizing committee, chaired by John, reflects how, by this time, prehistoric archaeology was firmly the domain of the Darwinists, who were keen to promote its importance to the wider debate on human evolution:

Lubbock	President of Organising Committee
	President Special Committee and Council
Huxley and Lyell	Member of Organising Committee
	Vice President Special Committee and Council
Pitt Rivers	Hon Sec. Organising Committee
	General Secretary Special Committee and Council
Spottiswoode	Treasurer Organising Committee
	Treasurer Special Committee and Council
Busk, Evans, Franks,	Member of Organising Committee
Hooker	Special Committee and Council Member
Prestwich	Special Committee Member

John was emerging as a natural leader of the "prehistoric movement" within this Darwinist fellowship (Van Riper, 1993). His collecting activity during this period reflects the linchpin role he played in the wider brotherhood.

By the time of his death in 1913, John had collected over 1,300 artefacts from nearly 400 individual 'transactions'. The nature of a transaction would vary; he would receive gifts, buy material from commercial dealers, and sometimes collect artefacts directly from archaeological sites. John's immediate circle of friends was a major source of items for his collection, accounting for sixty-five transactions in total. These were mostly gifts. Thirty-nine of these involved key members of the X Club and Lubbock-Evans network: Evans, Hooker, Busk, Franks and Pitt Rivers. These five compatriots sent John parcels by mail and perhaps brought artefacts in small cardboard boxes or cloth bags to social gatherings and society meetings. Here they would have been the focus of conversation for a while with a small group of people in the room.

Ever since their early interest in Scandinavian prehistory, Evans had been an important collecting companion for John. Evans was an avid collector of prehistoric artefacts and donated material to John on eleven separate occasions during the period, 1863-1881. When John and Nelly visited Evans' home at Nash Mills in June 1865, Evans probably gave him a selection of stone tools which he had discovered on the shores of Lough Neagh in Ireland.[109]

'[Inviting Lubbock to talk about a planned trip to Denmark] and to show you some more Lough Neagh things I found with the help of two boys 19 celts last Sunday at [Toone], many of them however mere fragments.'[110]

His other donations included a collection of stone tools from Spiennes in Belgium,[111] a 'Paraguay Indian horn',[112] and a Palaeolithic stone axe from South Africa.[113] Evans' pocket notebook covering the period 1866-1868 includes scribbled notes from his travels in Austria, England, France, Switzerland and Germany.[114] One of the small yellow-white pages contains the following entry written as a list:

> 'Bronze objects found at the Camery, Saxony.
> Lubbock 18/
> Palstaves
> Sickles
> Bracelet
> Sp. Head'

The entries 355-9 in John's catalogue of his collection record the acquisition of a winged axe and sickles from Camery, Saxony, 'bought for me by J. Evans 1866.'[115] In April 1876, Evans went to France and on 14 April visited the 'forest of rocks' at La Gauterie:

> 'Visited the Bois du Rocher at La Gauterie about 5 miles from Dinán near St. Helen. The implements are found on the slope of a hill facing the southwest and with a considerable [haul] of lowlying country in front. They occur in the surface soil in considerable abundance as I found myself, and probably are not confined to one spot…Most of the implements are of quartzite but some are of flint.'[116]

Evans gave an item he collected during this trip to John in June 1876[117] while John also contributed to Evans' collection. In 1868, he gave Evans three bronze implements he had bought in a small antiques shop in Naples during a trip in April of that year.[118] He also gave Evans three cores found in the Rohri Hills in the Sind region of India in 1876.[119] These had been sent to John from General Sir William Merewether in September 1875.[120]

Hooker presented John with gifts for his collection on ten occasions during the period 1867-1873. All the material he donated probably originated from his global network of contacts set up to develop the Museum of Economic Botany at the Royal Botanical Gardens, Kew:

'The Commissioners of Her Majesty's Works having been pleased to form a Museum of Economic Botany within the Royal Gardens, the director solicits the co-operation of Her Majesty's Ministers and Consuls in foreign countries, of the Governors of Her Majesty's Colonies, of Officers in the Army and Navy, Merchants and Travellers generally, to aid in contributing specimens towards so desirable an object.'[121]

In February 1867, Hooker gave John a North American Indian carving and pipe that had been brought to England by 'J. Douglas'; possibly Sir James Douglas who had been Governor of British Columbia and Vancouver Island until 1864 when he had retired and embarked on a tour of Europe.[122] During a six year period, Hooker sourced stone flakes from New Zealand[123] and from near King Williamstown in South Africa;[124] a 'walking stick' from the South Pacific Islands;[125] stone implements and pottery from the Rupununi River in British Guiana;[126] pipes, a horn and a fire stick from indigenous tribes in South Africa;[127] shell axes made in Barbados;[128] and an Indian ornament from the upper regions of the Amazon River brought to England by Mr Spence.[129]

Franks, Pitt Rivers and Busk between them donated artefacts to John on at least eighteen occasions during the period 1866-1880. Busk, for example, donated a hafted stone axe from the Amazon in August 1870,[130] which had been brought back from Rio de Janeiro by 'Mr. York'. In May 1876, he presented a selection of quartzite Palaeolithic implements discovered close to the large fallen capstone at the prehistoric site of La Gauterie.[131] Franks' donations included stone flakes found at Bethlehem in Israel collected by Mr Tyrwhitt Drake, a blowpipe from the Dayak people of interior Borneo, and prehistoric stone flakes found in drift gravel near Rome.[132] Pitt Rivers gave John a set of poisoned arrows and a quiver in March 1869, arrows from Demerara in British Guiana, a woman's lip stud from Central Africa, and in early 1868, prehistoric worked flint from his excavations at Cissbury Camp.[133]

John's collection was small in comparison with that of Evans and Pitt Rivers.

'I ought to have seen Lubbock's collection before I had seen yours. You have infinitely more and finer.'

(Canon Greenwell in Evans, 1943, p.124)

Evans and Pitt Rivers became obsessive and prolific collectors, however, John was not motivated quite to the same degree; perhaps because John came from a different collecting tradition to the antiquarian school where a collection was an end in itself. John's inspiration had been Darwin and his scientific training, where collections were made to provide evidence in support of a theory.

Yet all three collections were growing subsets of a much larger 'meta-collection' of prehistoric and ethnographic artefacts being created in and around London. Collecting was not a gentle form of leisure activity for any of them, but a vital part of the Darwinist agenda. Real artefacts from the distant past and the exotic present provided real evidence for evolution in just the same way as fossils, plants and zoological specimens. They could be rigorously assessed, compared and classified by identifying points of similarity and difference. Private collections were not locked away for the enjoyment of the owner and close friends; they were shared with the wider scientific community both at home and abroad, and actively used as a research tool. They also begun to be shared with and influence the collecting activities of the public museums.

Before the 1860s, the British Museum did acquire random prehistoric archaeological and ethnographic material often brought back to England by military expeditions and adventurous travellers. However, it struggled with how to interpret and display them, other than as exotic items of curiosity. All this was to change after 1865, when John, Franks and Hooker became trustees of the Henry Christy bequest. Christy had made his money through family business in the hatter trade and through his directorship of the London Joint-Stock Bank. Inspired by the Great Exhibition at Hyde Park in 1851, he spent the rest of his life engaged in ethnographic travel and research. Everywhere he went – Norway, Sweden, Denmark, Mexico and British Columbia – he collected prehistoric archaeological and ethnographic material. He met Thomsen, the father of the Three Age System, in Copenhagen in 1852, and in 1861 one of Thomsen's assistants was asked to catalogue Christy's collection according to these principles.

In 1858 he used his wealth to set up in partnership with his friend Edouard Lartet to excavate the Palaeolithic cave sites of the Vézère Valley in Southwest France. Whilst excavating, he caught a serious cold that combined with other complications to cause his premature death at fifty-five years of age. Christy's collection of approximately 1,000 items was bequeathed to the British Museum under a section of his Will dated 5 February 1863. The Trustees of his estate (John, Hooker and Franks) were empowered 'to make over the said Collections either in whole or in parts to an Institution'. Christy's Trustees transferred the collection to the British Museum in November 1865, although due to lack of space the collections remained at Christy's house, 103 Victoria Street, for several years. In his will, Christy had also provided £5,000 for the future development of his collection. Franks, as superintendent of the British and medieval antiquities department at the British Museum, used the interest on this amount, under the guidance of John and Hooker, to buy approximately 20,000 artefacts for the collection now held at the British Museum (Cook, 1997; King, 1997, p.140).

By the time Henry Morton Stanley found Dr David Livingstone near Lake Tanganyika in 1871, the Darwinist brothers in arms were strongly in the ascendancy within London intellectual society. They had their collective feet wedged firmly in the opening doors of the Royal Society, the BAAS, the Anthropological Institute, the Society of Antiquaries and, more tentatively, the British Museum. However, their influence on the London elite is only part of this story. Equally important to them was the winning of people's hearts and minds within the wider British community and the spreading of their gospel according to Darwin across the globe.

Chapter 5

The Bone-Caves
Implement

Carefully guided by John's steady hand, the steel pen tip slowly traced the fine lines of ink across the crisp woven paper, following the initial pencil pattern already laid down. Hunched over his desk, stopping every few seconds after each line, he turned to his microscope and checked that his drawing remained faithful to the original. He had learnt these close observational skills over a decade before from his master and mentor, Darwin. On this occasion, the subject of his attention was a black piece of cherty flint, roughly the size of his fist, shaped over 40,000 years ago by someone who had carefully removed flakes of flint to create a tool. It felt a good solid weight in his hand and the edges, though worn after so many years in the ground, still had a sharpness that hinted to their original function. This was a stone tool from the 'Old Stone Age' or 'Palaeolithic' period that would have been used by its original makers to cut, chop and scrape plants, wood, meat, hide or bone.

John would have recalled the excitement with which he had collected this artefact himself during his visit to the cave sites being excavated by Christy and Lartet in southwest France.[134] On 27 March 1864, the party had travelled along the Vézère Valley from Les Eyzies to a prehistoric rock shelter located in the limestone cliffs near Peyzac-le-Moustier on a large bend in the muddy brown river. At 'Le Moustier' he acquired a collection of 'rude implements'[135] perhaps discovered by him during the visit or given to him as parting gifts by Christy and Lartet. The stone implement John was now drawing proved to be the most interesting of this small group of material and is now housed at the British Museum in London.[136] It had been worked on both sides into an ovate shape with a point at one end. In John's eyes it was an enigma that he struggled to explain. It did not seem to fit comfortably with the early Stone Age axe 'drift forms' discovered by Boucher de Perthes in the Somme Valley, nor was it finely polished like some of the flint implements discovered in Scandinavia.

This specimen was so intriguing that a year after its acquisition John and his publishers decided that it warranted the expense of being included in his brand new publication, *Pre-historic Times*, in the way of three illustrations. One of over 100 wood-engraving firms in London at this time would have been commissioned to produce the wood blocks required. A skilled artisan created the reverse images of the supplied pen and ink drawings before carefully engraving three small squares of boxwood – the only timber with sufficient closeness and toughness of grain to create quality images in print. Figures 131-133 in the first edition depict both faces and a side profile of this important specimen (Plate II):

'MM. Christy and Lartet regard this type as identical with the "lancehead" implements found in the drift. I cannot altogether agree with them in this comparison. Not only are the Le Moustier specimens smaller, but the workmanship is different, being much less bold. Moreover, the flat surface (fig. 131 A) is no individual peculiarity. It is very frequently, not to say generally, present, and occurs also on the similar implement found by Mr Boyd Dawkins in the Hyaena-den at Wokey Hole…This very interesting type seems rather to be derived from the "cutters" above described, at the same time its resemblance to the drift forms is certainly great. MM. Christy and Lartet, indeed, call the implements of this type "lance-heads"; but it may well be doubted whether they were intended for use in this manner, though there are other specimens at Le Moustier which have all the appearance of having been intended for this purpose. On the whole, then, although these Le Moustier types are of great interest, we must pause before we regard them as belonging to the drift forms. No polished implements have yet been found in any of these caverns.'

(Lubbock, 1865, p.251-3)

Spreading the Gospel

John and his collections played a critical part in popularizing the human evolution and antiquity debate with the middle classes at home and abroad. His main scientific contribution was the publication of *Pre-historic Times*, a book that ran into seven English editions during his lifetime, which sold 20,000 copies (Sherratt, 1983) and was translated into several languages, including Danish and German. The most important English editions were the first and second, published in 1865 and 1869 respectively.

When Lyell published his disappointing '*The Geological Evidences of the Antiquity of Man, with remarks on theories of the origin of species by variation*' in 1863, John was encouraged to publish the various articles he had written to date in a single volume for wider consumption.

'Sir Charles Lyell has of course sent you his book. He devotes so little space to Scandinavia and Switzerland that I am thinking of expanding and reprinting my articles. Have you any young man who would give me, in English, an epitome of your recent researches & discoveries. If you know of anyone who would undertake this for me, I would with pleasure pay whatever you considered fair. I am aware that I am asking a great favour, but I am very anxious that your researches should be appreciated here.'[137]

John also felt a little aggrieved that Lyell had not, in his view, adequately acknowledged the recent work by himself, Prestwich and others regarding the drift stone implements.[138] He published a review of Lyell's book in the *Natural History Review* edition of April 1863. This review, though constructive and polite, expressed frustration at this 'unintentional' plagiarism of other people's work, and disappointment at the lukewarm acceptance of Darwin's theories of natural selection and variation in Lyell's book (Lubbock, 1863a).

John wanted to speak to the wider world directly rather than have his messages and ideas filtered through the work of others. He also had a company willing and ready to publish his work – the same publishers who two years before had started to print the *Natural History Review*. Williams & Norgate was a scientific and foreign book publishing house located at 14 Henrietta Street in Covent Garden, which was, at this time, the home of a thriving publishing community. Alexander Macmillan had opened a second branch of his Cambridge publishing house at number 23 in 1858 with Lord Tennyson, Huxley and Spencer on his books. John would have perhaps visited the offices of Williams & Norgate on his way to or from the City to discuss progress or deliver the precious completed handwritten draft manuscripts.

The first edition took two years to publish, appearing in print in May 1865. It brought together his earlier work on Scandinavian and Scottish køkkenmødding, the Swiss lake villages, the flint implements of the drift, North American archaeology, and the 'Cave-men' of southern France (Lubbock, 1861; 1862a; 1862b; 1863b; 1864). He expanded it with discussion on the use of bronze and stone in ancient times, the lives of 'modern savages' and megalithic monuments and tumuli. He connected all these thoughts and made sense of them within an overtly political and optimistic interpretation of Darwinism and natural selection, aspects of which he had first expressed publicly during his Royal Institution lectures on the antiquity of man in 1864:

> 'Thus, then the most sanguine hopes for the future are justified by the whole experience of the past. It is surely unreasonable to suppose that a process which has been going on for so many thousand years should have now suddenly ceased…The great principle of Natural Selection which in animals affects the body and seems to have little influence on the mind, in man affects the mind and has little influence on the body. In the first it tends mainly to the preservation of life; in the second to the improvement of the mind and consequently to the increase of happiness…
>
> 'The future happiness of our race, which poets hardly ventured to hope for, science boldly predicts. Utopia, which we have long looked upon as synonymous with an evidence impossibility, which we have ungratefully regarded as "too good to be true", turns out on the contrary to be the necessary consequence of natural laws, and once more we find that the simple truth exceeds the most brilliant flights of the imagination.'
>
> (Lubbock, 1865, pp.491-2)

Given the criticisms haunting Lyell's book, John was keen to ensure that *Pre-historic Times* was up-to-date with international research and gave due account to experts in the field. In the preface of the first edition, John lists the British and European public and private collections he had viewed, the sites visited and the people he had met 'in order to qualify' himself for undertaking such a publication.

> 'To Professor Steenstrup, Dr. Keller, M. Morlot and Professor Rüttimeyer, I am indebted for much information on the subject of their respective investigations. Finally, Mr. Busk, Mr. Evans, and Professor Tyndall have had the great kindness to read many of my proofs, and I am indebted to them for various valuable suggestions.'
>
> (Lubbock, 1865, p.x)

Ensuring this degree of accuracy was harder than he thought; and more difficult than we can perhaps imagine in our digital twenty-first century world where the immediacy of global connections is provided by email and the Internet.

One of the more unexpected challenges proved to be accessing information and illustrations from Denmark, not least because a war between two European countries got in the way. On 1 February 1864, Prussian forces crossed the border into Danish-controlled Schleswig. By early March, Austrian and Prussian forces had invaded Denmark itself. During the build up to war and for the next nine months, the attentions of Steenstrup, Worsaae and others were distracted from international archaeological matters. On 12 September 1863, John had written to Steenstrup asking for casts of engravings of Danish køkkenmødding material he wished to use as illustrations in his book.[139] He waited over three months for a response that eventually arrived in December. Steenstrup was required to obtain permission from the Danish Society of Antiquaries before consenting to the request. Now permission had been granted there was the question of how best to transport the images.

> 'Williams & Norgate say that the best way will be to take electrotypes (in copper and not white metal, as copper is much the best), and not to send the blocks, but the electrotypes to the Gyldendalsche Buchhandling in Copenhagen who are agents to Mr. W. + N. & will forward the parcel by the first opportunity, if you will kindly instruct them to do so. I should like to have all the figures used in your discussion with Worsaae about the division of the Stone Age.'[140]

The illustrations eventually arrived with John in April 1864. However, he was lucky to receive them at all! In May, Steenstrup had planned to send a paper to the Royal Society in England regarding his recent zoological research but:

'If I did so it might seem to you and your countrymen, that the feelings of the Danes for England and the English were the same as formerly. Such is not the case, and you could not expect, that so it should be. During the war the English diplomacy had not been the government of an amicable nation, nor of a neutral nation: and now not satisfied with a horrible war, Your Diplomacy is preparing a peace, more horrible…than the most horrible war!'[141]

This anger, though not directed individually at John or the scientific establishment, was typical of the disappointment felt by Denmark regarding the lack of support received from England against the Prussian invasion and the resulting loss of Schleswig to Germany.

The ability to illustrate and refer to prehistoric artefacts in *Pre-historic Times* was important to John for three reasons. In this book, he advocated a scientific approach to the study of prehistory informed by geology, and championed the division of prehistoric archaeology into '4 great epochs':

'Firstly that of the Drift; when man shared the possession of Europe with the Mammoth, the Cave bear, the woolly-haired rhinoceros, and other extinct animals. This we may call the "Palaeolithic" Period.

Secondly the later or polished Stone age…This we may call the "Neolithic" Period.

'Thirdly the Bronze age, in which bronze was used for arms and cutting instruments of all kinds

Fourthly the Iron age, in which that metal had superseded bronze for arms, axes, knives, etc.; bronze, however, still being in common use for ornaments.'

(Lubbock, 1865, pp.2-3)

To prove his theory, he applied a typology-based chronological framework to the prehistoric archaeological evidence available. Objects in his own collections and from elsewhere were used to assist in describing the type of object under discussion for each epoch.

'Figs. 66-69 represent small Danish flakes, forms exactly similar may be found in any country where the ancient inhabitants could obtain flint or obsidian.'

(Lubbock, 1865, pp.66-7)

He also drew on his collections, and those of others, to make a case for using the cultures of non-western communities living in other parts of the world during the nineteenth century as evidence to explore how people lived in Western Europe during prehistory.

'Fig. 156 represents the head of a Fuegian harpoon, which closely resembles the ancient Danish specimen figured in fig. 95.'

(Lubbock, 1865, p.436; Plate II)

John and his Darwinist friends believed that these hunter and gatherer communities who still used stone, wood and bone tools were remnants of civilizations living in Europe thousands of years ago. In their view of the world, Western Europe was at a more advanced stage of civilization than other countries, having evolved into the technologically, socially and culturally advanced nation states of the nineteenth century. Elsewhere in the world, society had stagnated and displayed forms of social and cultural development which Western Europe had already passed. Hence John's particular interest in the material culture of these communities and the inclusion of three chapters describing what he termed 'Modern Savages' in his book on the prehistoric archaeology of Western Europe.

Finally, John also referred to his and other collections of artefacts to emphasize his own direct involvement in discovery. He had been brought up in a culture of collecting for scientific purpose, inspired by Darwin, 'to collect merely for the sake of collecting, has a direct tendency to narrow the mind.' (Lubbock, 1855b, p.116). Specimens were there to be studied in detail so that their secrets could be revealed and provide evidence for larger theories of science. Those who studied these specimens and connected them into this bigger picture were important scientists in their own right, equal to those explorers who originally found the material. In the first edition of *Pre-historic Times* John makes thirty-two references to items in his own collection and the observations he had made about them.

The first edition was widely read by an intellectual public both at home and abroad. Darwin and his fellow X Club compatriots were complimentary (although Hooker was disappointed that John made reference to the charge that Lyell had plagiarized the work of others in 1863).

> 'The latter half of your book has been read aloud to me, & the style is so clear & easy (we both think it perfection) that I am now beginning at the beginning. I cannot resist telling you how excellently well in my opinion you have done the very interesting chapters on savage life. Though you have necessarily only compiled the materials the general result is most original. But I ought to keep the term original for your last chapter which has struck me as an admirable & profound discussion. It has quite delighted me for now the public will see what kind of man you are, which I am proud to think I discovered a dozen years ago.'[142]

It was also the last chapter, with its discussion of natural selection and implications for science, religion and human progress that fired up opposition to John and the book from those who supported the established church. John Phillips, Reader of Geology at Oxford University, gave the presidential address at the BAAS meeting at Birmingham in 1865. During his speech he outlined how science had progressed and mentioned the work on human antiquity. He referred to Lyell's 1863 publication, but made no reference to John or *Pre-historic Times*.

In 1865, John also put his name forward and was accepted as the Liberal candidate for the West Kent parliamentary constituency. As the railway extended its tentacles out into rural Kent in the early 1860s (Sevenoaks and Tonbridge), this picturesque corner of southern England became a fashionable place for London commuters to live. The eldest son from an influential local family that owned a bank in the City and regularly commuted into London should have had a good chance at contesting the election. However, the publication of *Pre-historic Times* three months before the election contributed significantly to what happened next:

> 'I do sincerely wish you all success in your election & in politics; but after reading this last chapter you must let me say Oh dear Oh dear Oh dear.'[143]

Darwin's written premonition in June sums up the inevitability of the election outcome. John had the option of publishing after the election but decided against such a course of action, wishing to be open and honest about his beliefs in public. The vote was held in July and John experienced, not for the last time, the public humiliation of what it was like to stand defeated on an election platform. He lost to Viscount Holmesdale and the Conservative politician Sir William Hart Dyke. Three years later, in 1868, he experienced the same fate again, stating 'I lost the West Kent election by 55. The Clergy were very bitter.'[144]

Yet John's work was far from ignored and it helped to shape the scientific, social, political and religious opinions of the 'middle classes'. Businessmen, government officers at home and abroad, career scientists and liberal members of the clergy all read and digested it. In 1869, he published a second and much enhanced edition of *Pre-historic Times*. In the four years between the first and second editions, John was engaged in a frenzy of national and international networking and collecting. We have already seen how he used his small and embryonic collection in the first edition (which included 32 references). In the second edition he referred to 55 items drawn from a mature and confident collection containing over 870 objects.

It is during the period 1865-69 that John's collection came of age and this reflects the role of *Pre-historic Times* as an international scientific gospel. If we plot the number of occasions when John collected artefacts during his lifetime against the year in which these acquisitions were made, there is a very definite 'spike' of activity coinciding with publication of the first and second editions of *Pre-historic Times*. Over forty-two percent were acquired during these five years, and a further twenty-one percent in the three years immediately afterwards. This intensity of collecting at the same point in his life when he is actively writing about human antiquity is no coincidence. John was conscientiously building up a personal scientific reference collection through purchase, fieldwork and donation from friends[145] to support his research. Correspondents from across the globe, who had read the first edition of *Pre-historic Times* with its reference to his collection, sent him artefacts to assist his work.

John also inherited the Lubbock family estate on the death of his father in June 1865. Although a deeply sad moment in his personal life, this event also transformed his bank balance and the persuasive influence of his chequebook. In 1865, John's income is recorded as £7,400 of which he spent £5,100 and saved £2,300. In 1866, the money coming in shot up to £19,600, of which he spent £14,200 and saved £5,400. His financial responsibilities would also have increased, but these are the accounts of an individual whose spending power dramatically changed after 1865. He could afford to travel more perhaps, to buy artefacts that caught his eye at home and abroad, and of course to commission the excavations at Hallstatt with Evans.[146]

A little excursion into northern France with Steenstrup in September 1865, a couple of months after his father died, is typical of the field collecting activity John undertook during his holidays over the next few years:

'John is still away. I expect him home on Saturday or Sunday. Just at present he is with Prof. Steenstrup in the country about Pressigny & Pont le Voye.

'He seems quite overwhelmed with the number of flints he has found - he says they filled two carriages in one day. I cannot imagine what it will cost to bring home his luggage, nor where we shall put the contents when they arrive.

'Believe Prof. Steenstrup will return with him, for the B. Ass. next week.'[147]

His trip to Hallstatt in 1866 with Evans has already been described. In September 1866, Canon Greenwell invited John to go barrow digging near Malton in Yorkshire:

'When I propose to open half a dozen or more very promising barrows... I need not say how glad I should be to enlist you as an ally in the cause of digging.'[148]

Greenwell was a canon of Durham Cathedral and a keen antiquarian who, during the 1860s, excavated a number of enigmatic prehistoric burial mounds ('barrows') dotted across the Yorkshire Wolds. Although regarded as a serious scholar in his time, Greenwell was of the 'antiquarian school' of archaeology. He rarely recorded the barrows he 'opened up' in detail, primarily focusing on digging down and extracting the artefacts and human remains buried inside. John and Nelly accepted his invitation, caught the train up to Weaverthorpe railway station in the Vale of Pickering, and stayed with Greenwell at the Pigeon Pie Hotel located on the High Street in Sherburn. On his return to High Elms John recorded three new entries in the catalogue of his collection; one for the flint flakes found by him on a tumulus near Sherburn, and the others noting gifts of 'Yorkshire flints' from Messrs Porter and Monkman who he met during his visit.[149]

In the spring of 1867, John spent a few days in Brittany with Hooker and Huxley, then travelled south into the Auvergne area of France with Evans and Franks. During this trip, he found a 'hatchet of flakes' on a tumulus at a megalithic site between Dol and Dinan in Brittany, purchased material in Lyons, Paris, Clermont Ferrand and Le Puy-en-Velay, and was presented with bronze implements by the continental archaeologists Gabriel de Mortillet, Dr Cussi and Monsieur Lacroix.[150] He took in a tour of Scotland, Shetland and the Orkneys after the BAAS meeting in Dundee later that year, collecting stone flakes from 'an Ancient dwelling' at Skaill and flakes made by him and his companions during an experiment to manufacture stone tools.[151] On 9 April 1868, he departed from London by train for a memorable trip to Italy:

> 'Picked up Tyndall at Amiens & Hamilton at Paris & with them & R. Birkbeck went to Naples to see the eruption of Vesuvius. We got to the edge of the crater twice while it was in eruption…We then went to Rome…& then Hamilton and I went on to Florence, Milan & over the St. Gothard to Zurich & Basle when he left me. I went on to Mayence, Dresden, Berlin, Schwerin, & Hanover to see the museums, & got back on the 9th May.'[152]

During this trip he acquired a considerable number of artefacts by both purchase and gift including a collection of bronze implements bought from a Naples antiques dealer, called Barone.[153] In May 1869, he went to Switzerland for three weeks with Nelly, 'going by way of Geneva, Neufchâtel, Thun, Grindelwald, Zurich & Basle'. Intriguingly, the scientific motivations for their trip were so that John could collect specimens of Thysanura (silverfish) and study new discoveries in the Lake Dwellings. His archaeological acquisitions on this occasion included axe heads from Lake Neufchâtel, presented by Monsieurs Desor and Coulon, and artefacts from the lake village at Robenhausen bought from Jakob Messikomer, the local archaeologist who had discovered and excavated the remains of six hut structures.[154]

Back home in London, John started to buy artefacts for his collection on a regular basis, particularly ethnographic material. On Wednesday 18 July 1866 at 3pm, the Arctic Collection of the late Richard Shingleton was sold in several lots at a small Croydon auction house owned by Messrs Crispe and Dracott. A few days earlier, John had obtained a copy of the catalogue, perhaps from the auctioneer's Bishopsgate Street office near where he worked at the bank. He may have visited the auction rooms on the previous day or on the morning of the sale itself to view the collection. He would have found an amazing array of natural history specimens, 'Esquimaux' dresses, bows, arrows and other artefacts collected by an Arctic traveller now long forgotten by history.

Richard Shingleton had ventured into the Arctic on at least two occasions as a bit player in an international drama that unfolded only a decade before. He served as the gunroom steward on the Royal Navy's HMS *Enterprise* when she set sail in 1850 in search of the missing Arctic explorer, Sir John Franklin. Franklin and his expedition had left England in 1845 on his final quest in search of the Northwest Passage, but all contact had been lost. Led by Captain Collinson, the HMS *Enterprise* expedition spent five years away from home searching and researching. However, they eventually returned without any news on Franklin's fate. Lady Franklin was not going to give up on her husband and, in 1857, she privately commissioned a crew to return to the Arctic on her yacht, the *Fox*. Shingleton was a member of the crew who went in search of Franklin again; on this occasion they did discover a note left ten years before by members of the original Franklin expedition in a cairn on King William Island. Dated 25 April 1848, it described how the expedition had been trapped in the ice for over a year and that twenty-four officers and crew, including John Franklin, were already dead. The remaining crew stated their intention to head south in search of rescue, but none of them were heard from again. On both the HMS *Enterprise* and *Fox* expeditions, Shingleton collected a formidable treasure trove of Arctic wonders. Seven years after his return from the treacherous frozen seas of the North his collection was put up for sale.

John annotated his copy of the sales catalogue in pencil with curly brackets, highlighting the lots he intended to bid for on the day of the auction.[155] Lots 16-21 included 'Esquimaux' pipes, harpoons, bone and ivory implements, small-scale models of canoes and a sledge, a pair of moccasins, and two bows with arrows stored in skin cases. The Avebury Catalogue entries 417-428 describe the purchase of at least some of this material, including the models, pipes, harpoons and ladles, on 18 July 1866. On the same day, a local Croydon gentleman named John Wickham Flower, who was a Fellow of the Geological Society and a collector with an interest in the antiquity of man, also presented John with an 'Esquimaux knife' from the Shingleton Collection.[156] Perhaps Flower was also at the sale. John's catalogue records two other instances during the late 1860s when he acquired objects at auction. In May 1869, he successfully bid for an eclectic mix of artefacts at Sotheby's: a bronze ring, South American lipstuds and a netsinker from the Lake District.[157]

Three months later his curiosity was roused and his wallet opened again by the prospect of buying a:

> 'Skull used as a drinking cup. Sold with Australian & New Zealand objects at the Marlborough House, Exeter, sale; & said to have been the property of Cap. Cook's officers.'[158]

John's other spending sprees during the late 1860s helped line the pockets of two London-based dealers in curiosities, both trading in exotica brought back from far away lands by travellers, explorers, colonial civil servants, mariners and soldiers. Wareham was a commercial dealer in Castle Street, just off Leicester Square, from whom Franks bought numerous prehistoric and medieval antiquities for the British Museum collections between 1860 and 1880.[159] John purchased ethnographic and archaeological artefacts from Wareham on twenty-nine separate occasions; seventeen of these fruitful visits to his shop took place during the critical five year period, 1865-9, between the publication of the first and second editions of *Pre-historic Times*.

The second shop that John frequented was located on the corner of Great Russell Street and Willoughby Street, within a stone's throw of the British Museum and at the heart of London's "natural history district". It was owned by Bryce McMurdo Wright Senior, a renowned minerals, shells and fossils dealer. He had moved from Liverpool to London in 1855 when trade in these commodities was increasingly profitable. He was a field collector himself but sourced most of his material from travellers returning from abroad eager to earn a few extra shillings from their endeavours. Some may even have collected commercially as a 'profession'. Anyone venturing inside the shop would have been greeted with a cocktail of sights and smells. John may well have visited Wright in the late 1850s and early 1860s to look at his natural history specimens, but it was a new business venture launched in 1868 that particularly caught his eye. The shop catalogue for that year included a single page list of "Stone and Bronze Implements" along with the following announcement:

> 'B.M.W. has just given orders to his correspondents (who are in all parts of the world) comprising Russia, France, Italy, Germany, China, India, North and South Americas, New Zealand etc. to purchase all Stone and Bronze Implements of whatever description, so as to meet the increasing demand made upon him for this new department of his business.'

(Cooper, 1999, p.13)

John's first purchase in January 1868 comprized of some flakes and arrowheads from Pennsylvania, North America.[160] In March of that year he bought a mixed collection of ethnographic material, including clubs from Fiji, which would presumably have been delivered to the estate at High Elms.[161] Of the twelve occasions when John bought material from Wright, eight of them were during the major collecting period of 1865-9.

John continued to exploit his connections with Scandinavia during the late 1860s, gathering new information and collections for his research. A single and unassuming entry in his catalogue during August 1867 hints at yet another significant purchase of prehistoric artefacts from Denmark:[162]

'549   ½ a collection made by Mr. Petersen. Bought with Flower through Engelhardt.'[163]

John had met Conrad Engelhardt on his trip to Denmark in 1863. At that time, Engelhardt had been an energetic thirty-eight year old director of the Museum of Northern Antiquities at Flensburg. The area was rich in prehistoric archaeology because of the exceptional preservation qualities of the local peat moorland. Engelhardt had carried out excavations at sites where rare wooden and other organic objects that normally decay had survived and been recovered. Perhaps his most famous discovery was the miraculously preserved Iron Age ship at Nydam in 1863 that inspired John to stop off at Flensburg during his visit in that year (Lubbock, E. 1864, p.369). However, Engelhardt's whole world turned upside down less than a year after John's visit. The same war that had helped delay the publication of *Pre-historic Times* in 1864-5 was to blame. The Second Schleswig War saw the important Danish port of Flensburg (second only to Copenhagen) become the property of the German kingdom of Prussia, along with the rest of Schleswig. In an act of 'heroic nationalism', Engelhardt smuggled the 'Danish' archaeological collections out of Flensburg into Danish territory, before he too fled to Copenhagen. These collections became the subject of a major international incident when Prussia demanded their return (Wiell, 1996). However, by 1867 the situation had calmed down to an extent and Engelhardt was firmly ensconced in his new role as a lecturer in Copenhagen and an assistant at the National Museum. It was in this guise that he acted as intermediary between John and the original owner of the collection recorded in the Avebury Catalogue as '549', 'Mr. Petersen'.

This enigmatic figure is likely to have been one of two candidates. Johan Christian Ludvig Petersen was a captain and school inspector whose son was

Henry Petersen, a respected Danish archaeologist. Johan also had a long-term interest in archaeology and had been collecting prehistoric artefacts since 1854. In 1887 he wrote to M.F. Herbst at the National Museum in Copenhagen to offer for sale 'a collection of stone age and bronze age artefacts' which had been valued at 9850 Danish Kroner (DKK) but which he would sell at a discount to the National Museum for 9000 DKK.[164] Could he also have sold a selection of material to John twenty years earlier? The other candidate is Julius Magnus Petersen. Engelhardt knew J.M. Petersen from his Flensburg days. He was a well-respected illustrator of antiquities who had drawn artefacts discovered at Nydam and other sites. He provided the illustrations for Engelhardt's recent publication about his peat moss discoveries: a publication which John helped to have published in English by Williams & Norgate in 1866. Petersen lived in Copenhagen and celebrated his fiftieth birthday in 1867. Perhaps he had a collection of prehistoric artefacts that he wanted to sell?

What we do know is that John at some point during his correspondence with Engelhardt offered to take charge of the disposal of 'Mr. Petersen's collection'. Given that 'Mr. Petersen does not speak English', Engelhardt arranged matters with John on his behalf. Petersen packed the artefacts in three boxes to be sent by steamer, and put prices on each piece. John shared them with Flower, a fellow collector with whom he had a mutual interest in the Shingleton Collection auction only a year earlier.[165] Engelhardt also included a few specimens from himself to John in the package, 'wrapped in yellow paper'.[166] One of the artefacts in this Petersen Collection, a partially polished stone axe head, was referred to by John in his second edition of *Pre-historic Times* and was the subject of two illustrations (figures 101-2). This specimen was used by John to demonstrate how the edges of these polished axes are sometimes re-sharpened, thereby showing signs of long use.

These were important items leaving Danish shores, and we have already seen how Engelhardt fought hard to save the Flensburg antiquities for Denmark a few years before. Perhaps part of the reason he was more prepared to let the Petersen material leave the country was because he regarded John and his work as important to Denmark. John's reputation internationally was greatly enhanced by the publication of *Pre-historic Times* in 1865. Over the next five years, the book was published in North America and translated into Danish, French and German. Strangers and acquaintances read the book, heard him talk at meetings and, noting his interest in collections and collecting, sent him gifts of stone tools and ethnographic objects from across the globe. Between 1865 and 1869, John's records show at least eighty-eight gifts to his collection. Through the stories behind these gifts we get a real sense of how *Pre-historic Times* inspired ordinary agents of empire overseas to take an interest in human evolution.

George Augustus Robinson had spent almost thirty years of his life in Australia motivated by a personal mission to protect and educate local aboriginal communities. He had returned to Europe in 1852 and in 1859 settled in Bath. In January 1866, the year of his death, Robinson sent John two pieces of wood of the type used as fire sticks by Tasmanian aborigines, the Palawa.[167] They had been given to him by 'his…friends' many years before and he had kept them as a souvenir of a community and culture now on the verge of extinction. Robinson suggested that John describe the Palawa process of fire-making in his second edition of 'your excellent work, *Pre-historic Times*', as he was afraid that otherwise the practice may go unrecognized by 'White Society'.[168] Robinson died a few months after writing this letter, but would have been proud to know that anyone reading the second and subsequent editions of John's work could see illustrations of these two pieces of wood accompanied by a brief commentary on fire-making in Tasmanian aboriginal society.

John received further gifts of Australasian material at this time. Mr Wood donated a stone axe from Queensland, and Mr Habgood some boomerangs, shields, spears and other implements from Western Australia.[169] Richard Birkbeck, a cousin of John's brother-in-law, donated another shield.[170] Mr B. Plant offered John a wooden bowl and two nets collected by a friend in the Australian interior, after reading *Pre-historic Times* 'with interest & pleasure'.[171] In April 1867, Monsieur Casolaui donated an African axe and bracelets from Khartoum, and in February 1869, Mr Dale presented a selection of flakes from the Cape Flats in South Africa.[172] Captain Richard Burton, the infamous and controversial explorer, donated 'Amazon bracelets' from the kingdom of Dahomey in West Africa.[173] Walter Elliot, an Indian civil servant and archaeologist, left a boomerang from Southern India for John at the Travellers Club in London on 29 June 1869.[174]

In November 1867, Eden Colville of the Hudson Bay Company donated arrows from the Hudson Bay Territory collected by a past Hudson Bay Company Governor, Sir George Simpson. Two years later he also presented a 'Hyda Indians pipe' found by him on Queen Charlotte's Island.[175] Paul Blackmore donated a small group of North American Indian artefacts collected by him from the northern shore of Lake Ontario in 1866.[176] His namesake, William Henry Blackmore, was a very wealthy English businessman who invested heavily and successfully in North America. He bought a vast collection of American archaeological artefacts excavated from prehistoric mounds in the Mississippi Valley, including the Squier-Davis collection which he purchased from Davis for $10,000. On 4 September 1867, he opened the Blackmore Museum in Salisbury to display this collection to the world. In August 1867, the Blackmore Museum presented John with a selection of North American 'Indian' and 'Esquimaux' artefacts.[177]

This dramatic acceleration in collecting activity, and the in depth networking with archaeologists and ethnographers that it reflects, fundamentally influenced John's second edition of *Pre-historic Times*. Of the seven editions published during his lifetime, it is the difference between the content of the first and second that is the most marked.

'I have endeavoured, as far as possible, to avoid unduly increasing the size of the book; and although the present work will be found to contain a great number of new facts, some of the chapters being indeed almost re-written, still it is only increased in size the extent of 100 pages. Nearly half of these are occupied by the addition of more than seventy new figures, which will tend to diminish, rather than increase, the time occupied by its perusal.'

(Lubbock, 1869, p.xi)

Despite this apparent concern that readers might think the longer book was a chore, the second edition reflected John's growing understanding of the subject matter and his interpretation of the evidence. His growing collection as a research resource helped to give him this confidence (Owen, 1999; Owen, 2000).

With the publication of a second edition, the gifts from far and wide continued to flow. One of his new correspondents, for example, was a twenty-eight year old surgeon and budding archaeologist, Charles C. Abbott, from Trenton, New Jersey. In a letter to John dated 6 June 1871, 'Chas' updated him on his own local fieldwork collecting activities and promised that he would present the whole collection to John for the purposes of 'science'. He had recruited several 'farmers' lads' to find ancient items lying scattered across the fields so that the collection would grow rapidly. He also planned to send a selection of unbroken specimens and a four-yard square sample of all material to John. Ploughing the ground uncovered the majority of these 'Indian relics' and he offered the ploughboys a few pence per dozen of unbroken or near complete specimens. He would send the collection once the ploughing had stopped:

'To the further increase of your museum; and the world's knowledge of what was once the population of the coast states of my country... if any specimen occurs worthy of a special remark I may have accorded to me the credit of having secured it unto science.'[178]

Abbott sent a collection of nearly 650 specimens across the Atlantic to High Elms. He enclosed a description and sketch of their find location in meadows on the New Jersey shore of the Delaware River, twenty-five miles north east of Philadelphia.[179] John recorded their arrival as catalogue entry 924 'a collection of stone arrowheads etc. Pd. By CC Abbott Esq.' and copied out Abbott's 8 June letter in full into the catalogue. He referenced Abbott's discoveries in subsequent editions of *Pre-historic Times* issued after 1871. Abbott himself was appointed an assistant curator at the Peabody Museum in Cambridge, Massachusetts, in 1876.

These numerous additions to his collection during the 1860s and into the early 1870s proved invaluable for the development of his ideas in *Pre-historic Times*, and he also used them to reinforce the evidence behind the arguments he made at learned society meetings. In 1863 he presented artefacts from the Swiss lake villages at the Royal Institution to illustrate his lecture on the topic. In 1870, he exhibited material at the Liverpool BAAS meeting. In 1871, he contributed stone implements to two exhibitions at the Society of Antiquaries on the caves and rock-shelters of southern France, and the Neolithic period. Other contributors included Evans, Lyell, Flower and Pitt Rivers. This showcasing of material was not just about John sharing the stories in his collection; it was also about confirming his personal credentials and those of the gospel he was proclaiming through his published work.

John was not writing *Pre-historic Times* in isolation. His book was one of at least four key texts available to buy in the late 1860s that spread the word about human antiquity. It sat on the bookshelves alongside the 1863 publications of Lyell and Huxley, and Edward Burnett Tylor's contribution entitled *Researches into the Early History of Mankind and the Development of Civilization* (also published in 1865).

'I have lately heard three Books, worth your attention—Lubbock Prehistoric Man— Tylor early History of Civilization, which is admirable; & Lecky's Rationalism, which also strikes me as very well worth reading.'[180]

Together they formed a pantheon of texts which were shortly to be joined by four further important publications in the early 1870s: John's own work *On the Origin of Civilisation* (1870), Tylor's *Primitive Culture* (1871), the culmination of Evans' ten year research project, *Ancient Stone Implements* (1872), and Darwin's very own *Descent of Man* (1871).

These books were designed to popularize Darwinist ideas, but ultimately their target audience was still the well-educated middle and ruling classes. Well into the 1870s, most 'lower middle' and 'working' class people in Britain believed in the Book of Genesis and the story of Adam and Eve. *Prehistoric Times* was never going to change that, with its use of French and other foreign languages including Latin without translation. John and Darwin of course were part of that well-educated upper echelon of society, but Huxley and Tyndall had very different upbringings. Huxley was a schoolteacher's son who was born above a butcher's shop (Desmond, 1994); Tyndall was a native Irishman whose father had been employed in the Irish constabulary (Barton, 1976). Through most of the 1860s, spreading the gospel to the workingmen of England had been as important to Huxley as convincing the members of London intellectual society. John joined him on a tour of the working class slum areas of Liverpool during the BAAS meeting of 1870, and gave a lecture to a group of 'Working Men' in Liverpool on 'the moral and social condition of savages'.[181] They became involved in the Working Men's Institutes, and championed the establishment of public museums that told stories of utopian progress through the use of objects and methods of communication that did not require the visitor to be able to read or write.

These were the actions of disciples who believed that the working classes could be inspired to evolve beyond their previously 'ordained' place in society through being converted to the religion of Darwinist science. Education was key to encouraging them to exploit new opportunities that would improve their condition in life. As well as demonstrating a clear sense of altruism, John shared with Huxley a passionate and overtly political view that Darwinist ideas about evolution and human antiquity could help answer wider social questions that ultimately underpinned late nineteenth-century British domestic and imperial ambitions.

Chapter 6

The Reindeer Antler Arrowheads

T he exhausted band of men sat huddled around the heat of the small fire they had managed to light whilst the inhospitable wrath of the Arctic seas closed in around them. Several hours earlier they had left Foggy Island off the northwest coast of Alaska and continued to travel westwards, determined to be the first European people to find a sea passage linking the great Atlantic and Pacific oceans from east to west. However, the relentless forces of polar nature were against them; they struggled as the weather, ice and fog closed in (Plate I). Less than fifteen miles down the coast they had been forced to find what little shelter this flat and barren landscape could provide. The height of the surf and shallowness of the seas thwarted their efforts to land in one bay they came across, which they christened 'Prudhoe Bay' on the maps of the coastline they drew. Tired and defeated, they had continued their seaward journey, battling the elements, until they found a gravel reef almost level with the water on which to camp.

Summoning up their remaining energy from deep within, they had dragged their boats ashore and collected the few willow branches available on their small refuge of dry land. The fire provided some welcome warmth and small protection against the onset of frostbite. Their naval 'uniform' was made of wool, flannel and broadcloth, and their boots of leather. It did nothing to protect their bodies from the freezing cold and wetness. Boots froze along with the feet inside them.

It was on this day, in August 1826, that their commander, John Franklin, told his men that he had made the difficult decision to give up their quest and turn back towards the east and home.[182] He was an experienced polar explorer and knew how quickly and dramatically an Arctic summer could end. The winds were blowing the sea ice in towards the shore and blocking their route along the coast. If they did not turn back now, the seas might freeze and prevent their return. In the next break of weather, they headed back to Foggy Island and began the year long journey back to family and friends in England.

Whilst Franklin and his men had been stranded on the tiny gravel island, a small British Royal Navy survey barge rounded Alaska's Icy Cape 160 miles to the west of Foggy Island. At longitude 71 degrees 23' 39" north and latitude 156 degrees 21' west they also met with the westerly gales that drove the ice down upon the coast. These storms forced the barge on to the shore and the crew spent the next few days trying to tow their heavy wooden boat across land back west towards their mother ship, HMS *Blossom*. Local Alaskan Eskimo people helped them and supplied venison and seal meat to eat in return for tobacco. When the winds eventually changed direction, Mr Elson, who was the master of the barge, and his crew were able to re-launch and sail down the Kotzebue Sound and rejoin HMS *Blossom* at Chamisso Island. It was no coincidence that he and Franklin were in the same area at an identical time. HMS *Blossom* was part of Franklin's expedition to discover an Arctic North West Passage that could be used as a trade route for British ships. Its role was to wait for Franklin in the West to collect him and his men when they were successful. Just 160 miles of gales, pack ice and fog prevented them from doing so (Beechey, 1831; Franklin, 1828).

However, HMS *Blossom* also had another mission to fulfil whilst she waited for Franklin; a scientific project which was much more fruitful. Franklin brought stories of adventure and exploration back to England in 1827 while Captain Frederick William Beechey and the crew of HMS *Blossom* brought back a large collection of artefacts from Alaskan Eskimo communities and detailed descriptions and illustrations of how they lived (Owen, 2006). Beechey was a student of Joseph Banks, the President of the Royal Society at this time and naturalist who had served with Captain Cook on his first voyage of discovery over fifty years before. Banks was a highly influential society figure with royal patronage; he and his friends persuaded the Admiralty to include scientific research and collecting in the mission of naval exploration across the globe. Collecting was an important scientific objective for the 1820s HMS *Blossom* expedition:

'As we have appointed Mr. Tradescant Lay as naturalist on the voyage, and some of your officers are acquainted with certain branches of natural history, it is expected that your visits to the numerous islands of the Pacific will afford the means of collecting rare and curious specimens in the several departments of this branch of science. You are cause it to be understood that two specimens, at least, of each article are to be reserved for the public museums; after which the naturalist and officers will be at liberty to collect for themselves.'

(Beechey, 1831, pp.viii-xv)

One of the young officers who took the liberty of collecting for himself was Lieutenant Edward Belcher. Born in Halifax, Nova Scotia, he had joined the Royal Navy at the age of thirteen. Now twenty-six, he was appointed to the HMS *Blossom* expedition as a surveyor. He was on the barge that had almost met with Franklin, and was an avid collector of Alaskan Eskimo artefacts. Collecting was not a difficult or dangerous occupation in these waters; the bartering of material goods was an important part of the Eskimo economy and they were used to exchanging items with 'western explorers' because of their long-term relationship with Russian travellers. HMS *Blossom* deliberately loaded up before departure from England with glass beads, jewellery, nails, saws and cloth as trade goods to exchange for food and other items. There were many opportunities to collect during their two summers in the Arctic:

'We hauled round the Western part of the Behrings…and tacked ship about two miles from some huts we perceived in a sheltered situation on the beach where the inhabitants were making preparations for visiting us. They launched four boats & came off, when a brisk barter soon commenced for Sea horse teeth, Bows & Arrows, fish spears, fishing lines, hooks & c., which they readily exchanged for glass beads or a few leaves of Tobacco.'

(Gough, 1973, p.145)

On their return to England in 1827, Beechey donated his collections to the British Museum and the Ashmolean Museum. Belcher donated a few items to the British Museum and the United Services Institute but kept most of the material himself. He appeared to have been less motivated by the idea of collegiate scientific discovery, and more by personal gain:

'About this time the Captain issued an order preventing the purchase of anything unless shown to him to ascertain whether he chose to detain it for Government (or truly himself) and after this always prevented any officer purchasing from the [umiaks] until he and his minions had completely weeded them of everything worth having.'

(Bockstoce, 1977, p.12)

It was not until 1860 that he first presented his HMS *Blossom* material in the public domain, with an accompanying paper, *On the Manufacture of Works of Art by the Esquimaux*, at the Ethnological Society of London (Belcher, 1861).

In March 1867, Belcher gave John three items from the Icy Cape which were probably collected during the HMS *Blossom* voyage – two 'Esquimaux arrowheads of reindeer horn' and a piece of 'reindeer sinew'.[183] Belcher provided a little further information in a letter dated 17 June:

'From within Pt. Hope in Lat. 68 degrees 19 N & Icy Cape 70.19 – others as far down as the Diomedes, Lat. 65.40…some "fancy specimens" – well finished…Aleutian Islands and are sent from thence, by the annual Russian [stow] ships to the Governors of their settlements on the western coast of America.'[184]

Collecting Artefacts, Acquiring Empire

John referred to Belcher's 1861 article in the first edition of *Pre-historic Times* on three occasions. It is tempting to conclude that Belcher had read the first edition and the references to his own publication, motivating him to donate a few artefacts from his own collection to John. It is also possible that the modern Eskimo arrowhead with the owner's mark, depicted for the first time in the second edition of *Pre-historic Times* (Plate III), was one of the two arrowheads given to John by Belcher in 1867 (Lubbock, 1869, p. xiii). In 1872 John acquired further items from Belcher's collection, probably at public auction, including Eskimo stone-tipped arrows, a spear with a stone head and a bone cup, all from Icy Cape.[185] These mysterious artefacts from an alien world, collected forty years previously by an earlier generation of British imperial explorers, were recycled into a collection now used by its owner, John, to support his own ideas regarding the late nineteenth century British Empire.

The handwritten catalogue entries recording John's acquisition of ethnographic objects, collected by European travellers from non-western communities across the globe, totalled at 426. He never travelled and collected ethnographic material first hand. Instead, he either bought these artefacts from dealers and auction houses or received them as gifts. He was an armchair ethnographer who relied on the journeys of others and avidly read of their exploits. He particularly collected objects from the communities of the Arctic polar regions; Australia, New Zealand and the South Pacific Islands; the African continent; and North America. Although some of the artefacts were made of stone, many of the ethnographic items he collected were made of wood, bone and other 'organic' materials that rarely survived in the European prehistoric archaeological record. Bows, arrows, clubs, axes, fishing equipment, harpoons, paddles, spears and carvings were his favoured types of object.

From the beginning, John's collection included both prehistoric archaeological and ethnographic material. The objects he purchased from Boye in 1863 included a representative selection of material from 'modern' Greenland communities as a comparison for the prehistoric artefacts. Boye provided a list of these objects as part of the English translation of his catalogue; he described how they were found in 'Esquimaux' graves in Greenland built of flat stones and in which the bodies were never burnt. Small sketch drawings of many artefacts accompany this list, including a 'harpoon of bone', a 'fishing hook made of two pieces' and a 'lamp of steatite'.[186]

About a year later, John was given a cheekstud by Dr Rae, another famous Arctic traveller and employee of the Hudson Bay Company who had also been involved in discovering the gruesome fate of the last Franklin expedition a few years before.[187] Rae had collected the stud from the local Inuit population of Mackenzie River in northern Canada. In 1870, he also gave John an Inuit harpoon head from Victoria Land, the coast of which he had surveyed for the Hudson Bay Company twenty years before.[188]

During the 1870s and 1880s, John increasingly received gifts of artefacts from officers, civil servants, missionaries and businessmen working in British colonies across the globe. James Hector, Director of the Colonial Survey and Museum in Wellington, New Zealand, wrote to John on 27 April 1870. It was a reply to John's earlier letter to Mr McKelvie of Auckland asking for some examples of early forms of stone implement. Hector promised to send John 'as complete a collection as I can afford from our duplicates'.[189] In September, John received 'stones and flakes' from New Zealand presented by 'Dr Hector'.[190] In July 1872, a letter was delivered to High Elms from Alfred Tozer of the Universal Mariners Insurance Company Ltd., 35 Cornhill, London, E.C. It gave John advance warning that he was being sent an adze handle originally collected in 1845 from the Maori of Middle Island, New Zealand by the explorer, naturalist and painter George French Angas.[191] Angas had illustrated this object in the published catalogue of '*Paintings Illustrative of Natives and Scenery of New Zealand and South Australia*', published in 1846. He had been Director of the Australian Museum in Sydney and his father had been instrumental in the colonization of South Australia during the 1830s. The stone adze head had been lost, but the handle entered John's collection in August 1872 as AC entry '1018'. Harry Cecil Cameron was a sheep farmer in New Zealand who returned to England in 1873, and on 27 November 1874 sent John a Maori implement he had found 'hoping you will accept'.[192] Not only was it accepted,[193] but also over 100 years later it now sits in the collections of Bromley Museum, London, as accession number 84.36.4.

In 1877, John William Brazier struck up a correspondence with John from his home at 11 Windmill Street, Sydney, Australia. His father was captain of a whaler. In the mid-1850s, father and son went on an expedition to islands in the South Pacific where junior became interested in natural history and shell collecting. In 1875, he had just returned from New Guinea as a member of the important Macleay expedition. Most of the specimens collected from this voyage are now in the Macleay Museum at the University of Sydney.[194] However, Brazier may have kept some aside because he offered John three adzes, one of flint from New Guinea (that he had found personally) and two of different stone from the small group of islands known as the Florida Islands in the Central Province of the Solomon Islands.

'I have read through the three editions of your *Pre-Historic Times* and therefore I think you are the right person to have the specimens I send.'[195]

Three years later, Brazier had moved further down Windmill Street to a 1850s terrace house, number 82. He sent a box of zoological specimens to Dr Gunther at the British Museum by the steamer, *Orient*, leaving on 23 January 1880. The package included a small parcel for John, addressed to him at the bank, 13 Lombard Street. It contained three ethnographic artefacts from Brazier's collection that he thought may be of use: a stone knife from New Zealand, an axe head from Port Moresby, New Guinea and an adzehead from Upolu on the Samoan Islands. In his letter he also corrected some misidentified artefacts in John's second edition of *Pre-historic Times*.[196]

By the time he sent his next gift to John, Brazier was the conchologist (mollusc shell expert) at the Australian Museum and was also in charge of the ethnological, historical and numismatic collections. The next gift was a large stone axe blade from New Guinea, which arrived in London by steamer, couriered by a Mr Goodwin who was also bringing a large collection of ethnographic material he had purchased in Sydney.[197] Brazier's following gift to John arrived on the steamship *Iberia* in 1886, accompanied by Mr Charles Harris, and was sent to his London address.[198] Eight specimens were donated, including knives and a scent box, which Brazier had obtained many years ago in the South Seas.[199] His final gift was made in November 1890, 'implements from New Guinea. Pres. By Mr. J. Brazier'.[200] In total Brazier supplied John with at least sixteen artefacts from the southern hemisphere for his collection.

The Avebury Catalogue entries 1025 – 1028 record John's acquisition of a bow and three arrows from the Indian tribes of eastern Brazil. The Portuguese who had colonized this area in the sixteenth century called these people the 'Botocudo'; the Portuguese word for 'plug' because of the distinctive wooden disks or plugs they wore in their lips and ears. By the nineteenth century these nomadic hunter-gatherers and their way of life were under considerable threat from increased European activity. It was a small, hard and glasslike mineral that brought this new, unwelcome colonial interest – diamonds had been discovered in the river deposits of eastern Brazil. European settlers deliberately spread smallpox through the local Indian population and left poisoned food scattered in the forests.[201] Today, a few tribes remain in rural villages and on Indian reservations and the artefacts of their ancestors live in museum collections across the western world. Chas Browne was a nineteenth century European traveller who spent a few years in the company of the Botocudo. He donated an example of their bow and arrows to John and also donated poisoned war arrows to Franks for the British Museum Christy Collection.[202]

Henry Bartle Edward Frere wrote to John from his desk at the India Office, Westminster, on 21 August 1874. The purpose of his letter was to provide further background information regarding a box of flints sent to John by Sir William Merewether on Frere's behalf. The flints came from the Sind region of the Indian subcontinent (now in Pakistan) where Merewether was Commissioner.

It was a part of the world Frere knew well. At the age of nineteen he had been selected as a writer for the Civil Service in Bombay (now Mumbai). In 1850 he had been appointed Chief Commissioner of Sind and in 1862 he became the Governor of Bombay. He was a career colonial administrator, temporarily based in London when writing to John. He had recently returned from Zanzibar where he had negotiated a treaty with the sultan, Barghash bin Said, regarding the suppression of traffic in enslaved people.

The worked flint nodules (nuclei) arrived in late August/early September 1874 and John recorded them in the Catalogue as entry '1078'. In his letter, Frere explained how they had been found in low limestone hills near the city of Sukkur through which the great River Indus flows. He first observed these strange shaped flints twenty-five years ago but assumed they were natural until he saw *Pre-historic Times* and realized they were of human manufacture. Merewether found these particular specimens whilst constructing canals near the area where Frere had first discovered them. Frere also provided John with Merewether's address in Karachi, and told him that he also sent a large number of specimens to the Geological Museum twenty-five years ago.[203]

John apparently wrote to Merewether in September shortly after the arrival of the flint package. However, Merewether did not reply from the 'Frontier Districts' of 'Upper Sind' until 15 March 1875. He had waited until he had the opportunity to revisit Sukkur and search the area Frere mentioned in his letter. However, he had been unable to find any specimens. Since then he had obtained a few more examples from the 'Executive Regencies' of the District, which he promised to send to John. He asked in return to be sent a copy of John's identifications 'if not too much trouble' so that he can use them to catalogue other material found in the Sind.[204] In September 1875, one year after the first batch, another package of Indian flints arrived at High Elms, and was processed into John's collection.[205]

Two years later, Sir Bartle Frere (recently knighted) was appointed as High Commissioner for South Africa and moved to Cape Town with his family. His wife, Lady Catherine Frere, wrote to John from Government House on 20 November 1877 to warn him to expect a 'curious baked earth bushman pot' that she had sent with Captain Penfold on behalf of her husband. It came from the 'Cold Bokkeveldt' in Worcester Province, and was already broken and mended before Frere received it. Penfold left it at the Royal Society for John to collect.[206] The earthen vessel arrived just in time for Christmas and was labelled '1148'.

This was not the only African artefact sent from South Africa to John by

British colonialists. In September 1873, Captain Caines departed Cape Town on the Union Steam Company's Royal Mail Steamer, *Briton*, destined for Southampton, England. Amongst his luggage, he carried a package of material for John sent by a Cape Town attorney, Charles Aken Fairbridge. Charles' parents, Sarah and John, had emigrated as newly weds in 1823 from Stepney, Middlesex, to bring up their family in a new life overseas. John Fairbridge became an established medical doctor in Cape Town and Charles was born a year after the couple had stepped ashore on the African continent. He became a respected legal attorney and Queen's proctor, who on his death left his collection of 7,000 books on the subject of South Africa to the South Africa Library (Mckenzie, 2005; Merrington, 1995).[207] Fairbridge sent John a selection of 'crude' stone implements and boulders used as grinding stones by 'Bushmen' (more properly called the San people of South Africa who lived by hunting and gathering). He also sent fragments of 'ancient native pottery from Cape Point' and a small grooved stone found near the Orange River, possibly used by 'bushmen' to sharpen bone arrowheads usually made of ostrich. One box contained a 'bushman' skull.[208] According to John's catalogue, he only acquired a single item from this delivery; the 'stone used by bushmen to sharpen arrows'.[209]

Back home in Deptford, southeast London, the reverend Julius Kessler sent John various items he had collected during his travels in Madagascar.[210] In his accompanying letter, he commented on the cultural similarities between the Madagascar peoples and the population of Fiji in the South Pacific. He had read John's published work and encouraged him to consider in further detail the curious nature of Madagascan traditions, which blended African, Asian and Arabian language and cultures.[211] In 1870, a London-based Gold Coast trader, Andrew Swanzy, donated stone implements to John's collection that had been found on the Gold Coast near Accra. Swanzy had funded an expedition into uncharted territory by the writer and explorer, William Winwood Reade, under the auspices of the Royal Geographical Society. During this trip, Reade had met a German missionary at Adumassie, Mr Zimmerman, who told him about stone implements discovered by local people in the gullies created after heavy rainfall. These rainstorms were usually accompanied by thunder and lightning, and the native people believed these stone tools to be 'fetish' thunderbolts, calling them 'god-axes'. When powdered they were believed to be good for treating rheumatism. Reade had bought them from the people who had discovered them. According to him there was no fear of deception as they were found so deep and the 'natives are far too lazy

Plate I

'The Beagle in the Murray Narrow, Tierra del Fuego'. Watercolour by Conrad Martens, artist on the HMS *Beagle* voyage with Charles Darwin. © *English Heritage (English Heritage Images ref. J980080)*

'Boats on a swell amongst ice'. Drawing by Captain George Back reproduced in the *Narrative of a Second Expedition to the Shores of The Polar Sea in the years 1825, 1826 and 1827* by Captain John Franklin (chapter 6). © *Scott Polar Research Institute, University of Cambridge*

Plate II

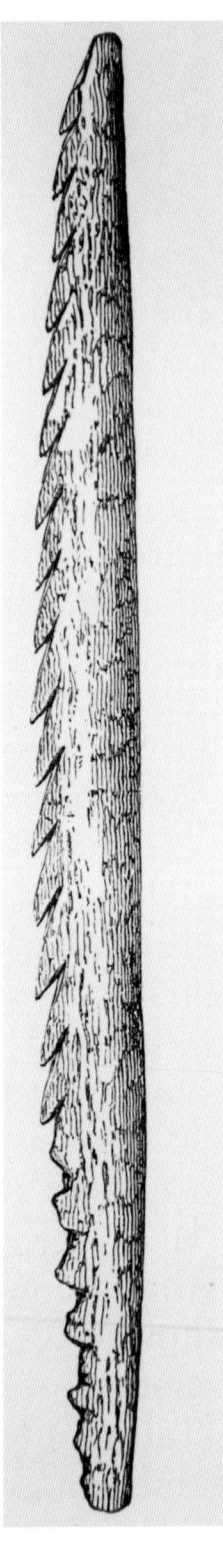

Illustrations of a flint implement
from Le Moustier which Lubbock
collected during a visit to the
Dordogne region in 1864 (chapter 5).
(Lubbock, *Prehistoric Times*, 1913, Figures 218-9,
p.324-5)

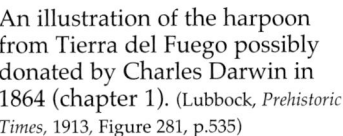

An illustration of the harpoon
from Tierra del Fuego possibly
donated by Charles Darwin in
1864 (chapter 1). (Lubbock, *Prehistoric
Times*, 1913, Figure 281, p.535)

Plate III

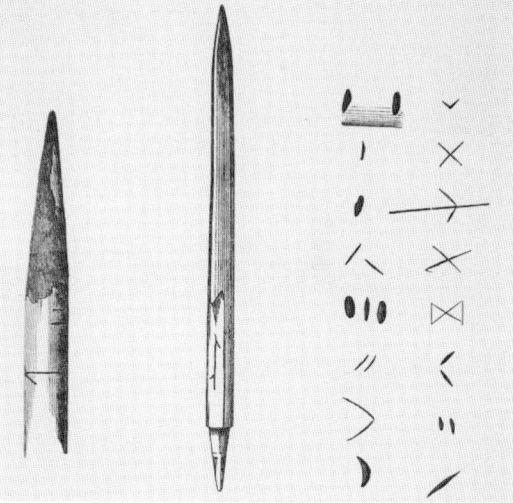

Extract from the Avebury Catalogue, Volume 1, written by Wilhelm Boye for John. It describes one of the tumuli in the field at So'mark on the island of Moen, Denmark from which Boye collected numerous prehistoric artefacts, including flint dagger '111' (chapter 2). *Courtesy: Bromley Museum Service*

Figures 1-3 from *Prehistoric Times* (Lubbock, 1913, p. 13). This image compares an ancient Danish arrowhead with owners marks, a modern Esquimaux example from John's collection and other owners marks from other ancient Danish arrows (chapter 6).

Neolithic polished axe numbered '303' in the Avebury Catalogue, Volume 1. From 'Magleby Stevas' in Sjaelland, Denmark, acquired in 1865 and now at Bromley Museum (accession no. 68.63.9) (chapter 9). *Courtesy: Bromley Museum Service*

Plate IV

Sir John Lubbock, aged
33 years old. From a
drawing by George
Richmond R.A., in 1867.
Courtesy: Lubbock Family

Ellen (Nelly) Lubbock,
aged 22 years old. From
a drawing in 1856, the
year of her marriage to
John (chapter 1).
Courtesy: Lubbock Family

Plate V

Sir Joseph Hooker in his study. Pencil drawing by Theodore Blake Wirgman, 1886 (chapter 4).
© *Kew Royal Botanic Gardens, Kew Art Collection*

'Recovery of our porter'. Image of the porter rescued by Johann Joseph Bennen and John Tyndall after falling down crevasse on Jungfrau in 1862, by Whymper. John Lubbock was assisting, outside of picture (chapter 3). © *Alpine Club Photo Library, London*

Down House Sandwalk (chapter 1). *Courtesy:* © *English Heritage (English Heritage Image ref. K000185)*

Plate VI

High Elms, 1880. *Courtesy: Lubbock Family*

John, Alice and family at front door of High Elms, 1892.
Courtesy: Lubbock Family

Plate VII

Great Stone at Avebury, Wiltshire, 1896. Photographed by Sir Benjamin Stone as part of the National Photographic Record and Survey (chapter 8). © *Victoria and Albert Museum, London (V&A Search the Collections ref. E.717-2001)*

Plate VIII

The Lubbock Memorial at St. Giles the Abbott Churchyard, Farnborough, Kent (chapter 8)

to grind stones down in the way in which these have evidently been ground.'[212] There was also no tradition in recent history of these types of stones being used as tools. John acquired them in June 1870[213] and possibly displayed them at the BAAS meeting held in Liverpool during September.

From the Gold Coast to Cape Town, Brazil to British Guiana, New Zealand to New South Wales, Madagascar to the Arctic, the world was delivered to John at Lombard Street and High Elms through his colonial networks. This world was one of difference and otherness, of strange artefacts and exotic tales, which John captured and reshaped to his own ends.

John also collected information in the same way he collected objects. He took advantage of the wealth of library resources available to him including the family book collection at High Elms as well as those of his scientific colleagues at home and abroad and the learned societies to which he belonged. He read widely the journals and travel narratives written by famous explorers and anonymous travellers, past and present. Perhaps inspired by Darwin,[214] he would carry a notebook or slips of paper with him so that when he read something of particular interest he could take brief notes and record the relevant page numbers allowing him to find the information again easily. For example, he noted that on 26 June 1863 he read part of Schoolcraft's *Indian Tribes*, on 19 July 1864 he read about Weddell's *Voyage Towards the South Pole*, and less than a month later he was reading Grey's *Travels in Northwest and Western Australia* followed by Captain Cook's *First Voyage*, volumes 2 and 3 (which are still in the Lubbock family library today). He read at least seventy-four published journals and travel narratives during the period 1863-5,[215] including Darwin's copy of volume 1 of the *Voyage of the Beagle*, which he borrowed in September 1864.[216] His 'Notes on Savages' continues the theme,[217] recording his reading on this subject from 1867 onwards. He read seventy-nine further ethnographic references by 1870, including Franklin's *Journey to the Polar Seas*, Brett's *Indian Tribes of Guiana* (which he read on 13 December 1868) and Raffles' *History of Java*. There is a third notebook entitled 'Origin of Civilisation' with very similar content.[218]

Here, then, was his palette of 'ingredients' which he blended to paint his picture of Utopian progress; supplemented by conversations at lectures and scientific meetings, at private dinner parties, and in personal moments with friends like Hooker and Darwin who had themselves travelled the globe. It is a picture rightly uncomfortable to the morals, values and understanding of our twenty-first century eyes but is an important part of our story:

'If we wish to clearly understand the antiquities of Europe, we must compare them with the rude implements and weapons still, or until lately, used by the savage races in other parts of the world. In fact, the Van Diemaner and South American are to the antiquary what the opossum and the sloth are to the geologist.'

(Lubbock, 1913, p.430)

In the three chapters on 'Modern Savages' in *Pre-historic Times*, John described his view that they are not the 'miserable remnants of nations once more civilized' as once believed, but stalled at earlier stages of evolution which, by inference, can be encouraged to develop through Western assistance. In the first chapter, he described the Hottentots (modern name, Khoekhoe) and Bushmen (San), the Veddas that inhabited the interior of 'Ceylon' (now Sri Lanka), the Andaman Islanders, Australian and Tasmanian aborigines, the Fijian Islanders, New Zealand Maori and Tahitian Polynesians of the South Pacific. He talked about what they ate, wore and believed in, how they made their home, their languages, how they survived, their families and friendships.

He referred to the objects they used and the materials they had at their command. He made judgements about the sophistication of their behaviour based on the accounts of those who had seen it first-hand which were heavily influenced by his western view of life. The 'Hottentots' were 'the filthiest animals' because they were covered in grease, their clothes never washed, and their hair loaded with soot and fat (Lubbock, 1913, p.433). Although he did acknowledge that a number of travellers had commented on their integrity, friendliness and benevolence. The 'Bushmen' were, to him, even less civilized, having no knowledge of metals, no domestic animals, and no canoes (Lubbock, 1913, p.437). The inhabitants of 'Van Diemen's Land' (Tasmania) 'belonged to quite a different race, but were just as wretched as those of Australia' (Lubbock, 1913, p.453). As far as John was concerned, the Fijians had little regard for human life, with infanticide, human sacrifice and cannibalism commonplace.

'Yet amid so much that is horrible, there is still something in the Fijian which redeems his character from utter atrocity. If he hates deeply, he also loves truly; if his revenge never dies, his fidelity and loyalty are strong and enduring.'

(Lubbock, 1913, p.463)

THE REINDEER ANTLER ARROWHEADS

Tahiti was of particular interest to John:

'Tahiti, the queen of islands...may be taken as representing the highest stage in civilization to which man has in any country raised himself before the discovery of introduction of metallic implements. It is not, indeed, at all probable that any inhabitants of the great continents were so far advanced in civilization during their Stone Age.'

(Lubbock, 1913, p.471)

John read the accounts of Captain James Cook which told of how, when first discovered, the Tahitians had no knowledge of metals. Cook and his crew introduced them to iron and in a very short space of time the earlier weapons and tools of stone, bone, shell and wood had been almost entirely replaced. Here was a Darwinian example of technological and cultural evolution in action and recorded within living memory.

The opening scene of the second chapter on 'Modern Savages' presented the 'Esquimaux' (people known today by their indigenous names including the Inuit of eastern Canada and the Inupiat or Eskimo of northwest Alaska). These Arctic communities also had a special place in his work. Nearly twenty-three pages are dedicated to their story (Lubbock, 1913); the next best-represented group is the Tahitians with nineteen pages, followed by the North American Indians with thirteen. The 'Esquimaux' section includes ten illustrations compared to twelve illustrations in total between the rest of the ethnographic sections. Of the fourteen ethnographic items from his own collection referred to in the second edition, seven are both referenced and illustrated within the 'Esquimaux' section. Given the cost of illustration, this investment is a sign of their value and importance to John in presenting his case. He compared the owner's mark on a modern Eskimo arrowhead to the marked Iron Age arrows discovered at Nydam in Denmark by Engelhardt (Plate III). The engravings of Inuit and Eskimo stone arrowheads, knives and spearheads within John's collection, which were manufactured by a pressure-flaking technique, appeared to be almost identical to the prehistoric stone artefacts discovered across Europe, including the caves of southern France.

'In the preparation of skins the Esquimaux use certain stone implements (figures 114-116), which have frequently been overlooked on account of their simplicity, but which yet are interesting because they are exactly similar to certain ancient implements which are common in various parts of Europe, and have

85

already been described in page 94...The true nature and use of the ancient skin-scrapers has...been entirely explained by these modern specimens, with which they are absolutely identical.'

(Lubbock, 1913, p.504-5)

When talking about Christy and Lartet's excavations in Southern France, John commented on how the archaeological and geological evidence discovered suggested that his Western European ancestors lived in similar Arctic tundra climates to those now occupied by the Eskimo and Inuit:

'So far, then, as the present evidence relating to the Dordogne caves is concerned, it appears to indicate a race of men living almost as some of the Esquimaux do now.'

(Lubbock, 1913, p.331)

Although John detailed how they are 'excessively dirty' and 'great thieves', he finished with recounting Dr Rae's high opinion of them, adding:

'They seem from all accounts to present the remarkable phenomenon of a really high state of morality without anything which can be called religion.'

(Lubbock, 1913, p.512)

John drew heavily on Schoolcraft and other firsthand accounts to describe the North American Indians, from the 'semi-agricultural nations of the west' to 'the filthy barbarians of northern California'. The two illustrations reserved for this section portray methods of obtaining fire using a bow drill, and John commented that:

'Left to themselves they would perhaps have developed an indigenous civilization, but for ours they are unfit. Unable to compete with Europeans as equals, and too proud to work as inferiors.'

(Lubbock, 1913, p.525)

The final subjects of the second chapter were the South American Paraguay and Patagonian Indians, and the inhabitants of Tierra del Fuego that Darwin had met many years before:[219]

'If not the lowest, the Fuegians certainly appear to be among the most miserable specimens of the human race. The conditions of

their existence are very unfavourable, and their habits are of
special interest from their similarity to those of the ancient
Danish shell-mound builders, who, however, were in some
respects rather more advanced, being acquainted with the art of
making pottery.'

(Lubbock, 1913, p.537)

In the third chapter on 'Modern Savages', John used the descriptive
information about the societies to develop his Darwin-inspired theory about
cultural evolution. He started by commenting on the undoubted skill and
dexterity involved in designing, making and using the hunting weapons of
stone, bone and wood which he had just described. He suggested that
although the use of stone as the principal material of implements and
weapons characterized a society at an early stage of civilization, the nature
of this stage was not universal and therefore must be further subdivided. For
example, the Andaman Islander or the Australian aborigine cannot be
compared with the semi-civilized nature of the Tahitians. In the same way,
the cave dwellers of Southern France, with no domestic animals or
knowledge of pottery, cannot be compared with the Danish shell-mound
builders who had dogs or the Swiss lake village dwellers with their
knowledge of agriculture and weaving.

From his study of evidence, John also concluded that although similar
in form, many of the simpler weapons and implements were invented
independently by different societies across the world. Their similar designs
were driven by similar needs and raw materials; however, they showed
difference in the detailing. These societies also showed great variation in
their 'habits and customs', adding to the argument for independent
development. However, they also presented remarkable similarities, which
John used to prove the 'unity of the human race'. He compared their
approach to domesticating animals, making fire, burying their dead, family
relations, attitudes towards death, the sound of language, chastity and virtue,
clothing and respect, marriage and morals, the treatment of women, and
their religious beliefs. Then he drew upon all the 'scientific' ethnographic,
archaeological and geological evidence laid out in the book to put 'modern
savages' firmly in their place vis-à-vis the 'civilization' of Western Europe:

'Savages may be likened to children…the life of each individual is
an epitome of the history of the [human] race, and the gradual
development of the child illustrates that of the species.

87

'Savages have the character of children with the passions and strength of men...after making every possible allowance for savages, it must I think be admitted that they are inferior, morally as well as in other respects, to the more civilized races.'

(Lubbock, 1913, p.563)

In his final chapter of *Pre-historic Times*, John finished with a flourish of rhetoric. He drew on the principles of science and natural selection to argue for the human race as a single species that had branched out into groups, now at different stages of civilization. He stated his firm belief that continued progress towards civilization improved people's happiness and quality of life, that western society was on the threshold of civilization and how science and religion together could take it forward, moving beyond the confines of selection by nature to selection by man. He also noted how the life of a 'modern savage' remained one of poverty and struggle without a civilizing hand:

'Even in our own time we may hope to see some improvement; but the unselfish mind will find its highest gratification in the belief that, whatever may be the case with ourselves, our descendants will understand many things which are hidden from us now, will better appreciate the beautiful world in which we live, avoid much of that suffering to which we are subject, enjoy many blessings of which we are not yet worthy, and escape many of those temptations which we deplore but cannot wholly resist.'

(Lubbock, 1913, p. 594)

Interestingly, John commissioned Ernest Griset[220] to create a series of nineteen paintings during the late 1860s that depicted prehistoric and ethnographic life.[221] These works were clearly influenced by John's views on progress and inspired by the stories in his collection. They reflect our ancestors and their modern representatives as resourceful, brave and capable (Murray, 2009) and provide a unique insight into how John imagined the people and societies who had created the stone implements in his care. The Griset paintings were never published and remained a private resource for family and visitors to High Elms.

Shortly after *Pre-historic Times* was published in 1865, John started preparing a series of controversial lectures that expanded on this theme of

THE REINDEER ANTLER ARROWHEADS

'modern savages'. They were delivered at the BAAS meeting in Dundee and the Royal Institution in 1868 where they were met with support from his Darwinist friends and condemnation on religious grounds from the Church and the Duke of Argyll. Two years later they were published as a book entitled *On the Origin of Civilisation and Primitive Condition of Man* (Lubbock, 1870). The similarity in title with Darwin's work eleven years earlier is unlikely to be a coincidence. In the introduction, John gave four reasons for the importance of studying this subject. It shed light on the reader's own ancestors, helped them to understand themselves and their own customs, and provided an insight into their descendants' future. His fourth reason related to empire:

'The study of savage life is, moreover, as I have already observed, of peculiar importance to us, forming as we do, part of a great empire, with colonies in every part of the world, and fellow-citizens in many stages of civilisation.'

(Lubbock, 1902, p.5)

His concern was to ensure that Britain understood the cultures and traditions of its 'fellow-citizens' so informed political decisions could be made across the empire. Drawing upon his extensive reading, he charted a course through the subjects of 'art and ornament', 'marriage and relationships', 'religion', 'characters and morals', 'language' and 'laws'. He concluded by referring his readership to *Pre-historic Times* and restated categorically that science demonstrates how existing 'savages' are not the descendants of civilized ancestors, that the primitive condition of all men was one of 'utter barbarism', and that from this condition various races have independently raised themselves:

'If the past has been one of progress, we may fairly hope that the future will be so too; that the blessings of civilisation will not only be extended to other countries and other nations, but that even in our own land they will be rendered more general and more equable.'

(Lubbock, 1902, p.507)

The next two chapters will explore how John spent the next forty years of his life personally translating this theory into practice in his political work at home. First, we will take a brief look at how it might also have inspired his readership in the colonies across the British Empire.

The latter half of the nineteenth century witnessed a massive period of imperial expansion driven by the idea of free trade, with Britain as the world leader in terms of technology and economic power. The Industrial Revolution had begun in its heartland and had drawn upon the resources available across its empire. Networks of supply and demand between home and empire were well established by 1815 and gathered pace as the century progressed. Financial investment in the colonies became increasingly important to the British economy from the 1870s (Porter, 1987). As the owner of a London City bank, John would have been at the hub of this activity. By the end of the century, the white ruling class owned eighty-five percent of the earth's land surface, and the British Empire accounted for approximately twenty percent (Gash, 1978; McClintock, 1995). Britain was not alone, however, as other emerging European powers took their imperial stake, namely France, Germany and Russia. At first, late nineteenth century British expansion was about opening up new markets and exploiting new sources of raw materials. Eventually it was driven by the need to protect existing imperial trade links from growing competition, including the Suez Canal.

As far as John and his 'liberal' progressive colleagues in business, politics and science were concerned, imperialism was a natural outcome of self-help and progress towards a utopian civilization. His books and collection underpinned this argument with 'scientific fact'. The thousands of colonial administrators, military officers, doctors and missionaries crisscrossing the world on steamships and trains had plenty of time to read. *Pre-historic Times* and *Origin of Civilisation* would have sat on their bookshelves and in their travel cases alongside Darwin's *Origin* and *The Descent of Man* (1871), Huxley's *Man's Place in Nature*, Tylor's *Primitive Culture* (1871) and perhaps Evans's *Ancient Stone Implements* (1872).

These volumes may also have shared shelf space with accounts of a debate in the 1840s between two intellectual thinkers of Victorian Britain, Thomas Carlyle and John Stuart Mill. Carlyle's deliberately provocative perspective on 'modern savages' was that they were members of an inferior race that would never integrate into Western society and could be exterminated if they blocked the inevitable progress of (Western) civilisation (Bowler, 1993; Carlyle, 1849; Stocking, 1987). Mill responded with a liberal challenge, regarding all ethnographic and European communities as part of the same human family. Native peoples living within the colonies could be re-educated to integrate them into Western society, thereby ensuring their individual happiness and active contribution to the generation of wealth for the empire.[222]

The work of John and many other Darwinists followed the liberal approach advocated by Mill, a viewpoint that could be argued in one sense as more sympathetic to their 'modern savage'.[223] However, all of this work perpetuated an important myth of British white superiority over the rest of the world, including continental Europe. As the birthplace of the Industrial Revolution and the economic, technological and social wonders that this had made possible, it was Britain's perceived duty to share the wonders of civilization with the wider world. There was now a scientific rationale behind the idea of British superiority that could be interpreted as absolute fact. What was more, Britain could provide a shortcut, taking 'savage' communities towards Western levels of achievement and progress. Indeed there was a sense that it had a moral duty to do so. Rudyard Kipling labelled this the 'white man's burden' (Kipling, 1899) and it influenced government policy and action on the ground.

For one example of how these ideas filtered into practice, we return to a member of John's colonial network. In 1867, a small girl discovered a diamond near a small British trading outpost called Cape Town in southern Africa, strategically located on the main sea route to India (McClintock, 1995). Ten years later, the Colonial Secretary Lord Carnarvon appointed Sir Bartle Frere as High Commissioner of South Africa. He arrived in a Cape Town transformed by the influx of diamond hunters and settlers, with instructions to create a confederation of British South Africa and the neighbouring Boer State and Kingdom of Zululand. Disraeli had given explicit instructions that Britain was to avoid a war with the Zulu, but Frere had different plans. From his view on the ground, he felt it was vital to break the power of the Zulu King Cetshwayo whose army threatened the recent British acquisition of Natal. The official story outlines how Frere sent an ultimatum to Cetshwayo calling for him to immediately disband his army and accept British 'suzerainty'. It was an ultimatum that Frere knew Cetshwayo could not accept as it meant accepting British rule. In January 1879, Frere commanded Lord Chelmsford to march on the Zulu army with 5,000 men. So began the infamous Zulu War at the Battle of Isandhlwana where 20,000 Zulu defeated an army of less than 2,000 British soldiers, leaving only 400 alive.

The Victorian novelist and traveller, Anthony Trollope, dined with Frere when he visited South Africa in 1877. He published a popular account of his travels that was reprinted in subsequent years. Three months after the start of the Zulu War, in April 1879, Trollope wrote to an Australian colonial administrator and historian, George William Rusden, strongly hinting that motives other than defence played a part in Frere's conduct:

> 'I cannot tell you how much to blame I think we have been in attacking Cetywayo. Frere, for whom personally I have both respect and regard, is a man who thinks it is England's duty to carry English civilization and English Christianity among all the Savages. Consequently, having the chance, he has waged war against these unfortunates – who having lived side by side with us in Natal for 25 years without ever having raised a hand against us! The consequence is that we have already slaughtered 10,000 of them, and rejoice in having done so. To me it seems like civilization gone mad!'
>
> (Hall & Burgis, 1983, p.826)

He also expressed these views in a later edition of *South Africa*, which prompted Frere to send a 100 page manuscript in response.

We know that Frere was familiar with John's work and motivated enough by it to send him artefacts for his collection. In some small way, it is possible to argue that the ideas expressed in *Pre-historic Times* and *Origin of Civilisation*, underpinned by John's collection, contributed to the tragic Zulu War of 1879.

Chapter 7

The Honduran Spearhead

John forgot about his long journey and aching limbs as the carriage swept through the gates into the magnificent grounds of Knowsley Hall. This grand estate near Liverpool was owned by one of the richest families in England. Over 100 years ago the great English garden architect, Capability Brown, had been commissioned to create one of his masterpiece landscapes. John would have admired the fruits of this labour and been suitably impressed by the spectacular natural vistas, lake views and water gardens that revealed themselves as his coachman drove towards the substantial Georgian house. Visiting in October 1883 ensured John saw the Park and its broad-leaved trees at their rust-toned autumnal best. Finally, the carriage and its weary traveller came to a stop in front of a fine red sandstone and brick façade adorned with Doric columns.

Waiting on the steps to greet John was his host and new political ally, Edward Henry Stanley, the 15th Earl of Derby. Stanley had defected to the Liberal Party from the Conservatives three years earlier, having fallen out with Disraeli about foreign policy. He was a seasoned politician who had served as Colonial Secretary and Foreign Secretary on three separate occasions during the previous thirty-five years; twice appointed by his father, the 14th Earl of Derby, when Prime Minister.

The interior at Knowsley was designed to impress and flaunt the owners' pedigree and wealth just as much as the exterior. The entrance hall was panelled in carved oak and large paintings hung up the Grand Staircase. The State Dining Room was over thirty feet high and contained two Gothic fireplaces and a rare gilt bronze, or ormolu, chandelier. Lord Derby would have proudly taken his dinner guest on a tour of the house, as he had done for visitors so many times before; pointing out the portraits of his ancestors and showing John the library of natural history books originally collected by the 13th Earl of Derby. He might have told John more about how the natural history specimens collected by his grandfather formed the origins of Liverpool Museum. This interest in science and collecting had rubbed off on the grandson, and the 15th Earl also had his own selection of prehistoric archaeological and ethnographic objects.

It is tempting to think that John saw at least some of Lord Derby's collection during this visit, and that in amongst the politics they discussed a few of the artefacts and stories behind their acquisition. A Honduran stone spearhead may have been one of the items that caught their attention on this October evening, perhaps because of a detail or the quality of craftsmanship. John may have asked Lord Derby how he had come by it and would undoubtedly have been interested in hearing over dinner about his visits to the West Indies and South America during the late 1840s, a time when the United States, Britain and Nicaragua found themselves in dispute over a narrow strip of land known as the 'Miskito Coast'.

On 24 January 1848, James W. Marshall had been constructing a lumber mill in California when he discovered traces of a shiny metal that proved to be gold. So began the Californian Gold Rush and the US entrepreneur, Cornelius Vanderbilt, quickly saw a business opportunity in building a Central American canal that connected the Pacific and the Atlantic oceans. In 1849, he signed a contract with the Nicaraguan government giving him exclusive rights to build a canal following a route along the San Juan River and across Lake Nicaragua before cutting twelve miles through the Rivas isthmus to reach the Pacific. He also secured sole administration of a temporary overland trade route along the same line before the canal was built. The canal was never constructed because a preferred route further south in Panama was chosen later in the nineteenth century. However, in 1848 the inter-oceanic Nicaragua Canal was a definite contender and high on the imperial agenda for Britain and the United States. The temporary overland route was also proving lucrative with over 2,000 passengers per month paying $300 each during the first few years of operation. Britain wasted no time once gold had been discovered to ensure it had strategic control of the port called Greytown at the mouth of the San Juan River along the Miskito Coast. It also took a larger interest in territory further up the coast to the north, near the Belize River, where British loggers already harvested mahogany trees to supply furniture and cabinetmakers back home. In 1862, this land was declared as the colony of British Honduras in order to give Britain a foothold in any future canal development involving this geographical area.

Lord Derby, as a visitor to the region and an influential foreign policy decision maker throughout this period, knew this Central American story well.

Several years after John's visit to Knowsley, Lord Derby presented him with a Honduran stone spearhead, sourced from this politically sensitive area, accompanied by a selection of Japanese arrowheads, Palaeolithic implements from Amiens, and bronze objects from the Swiss lake dwellings.[224]

'I told Mr. Bryce Wright to show you a small collection of arms and implements which I began years ago, and never continued.
Of these I offered a choice to the Museum [British], and they were as you know accepted. I wish the rest to be in the hands of someone who understands and appreciates such objects, and I therefore asked him to offer you as many as you cared to keep.

'I hear that there are about 50 which will suit your collection, and I beg you to accept them, and as many of the rest as you may like to have.

You will really do me a favour, for I am unwilling to sell what I have collected, and have determined not to add to its number. They cannot be in better keeping.'[225]

Taking Over the Asylum

When John visited Lord Derby a few months before his fiftieth birthday in 1883, he had arrived both at Knowsley Hall and at the pinnacle of his career. For the last fourteen years John had served as a Member of Parliament for the Liberal Party, and become a powerful and highly respected member of political society as well as a distinguished banker and scientist. Like many of his Darwinist friends, John was no longer a young radical brother in arms but had taken over the asylum and was now very much part of the establishment.

Lord Derby was a political ally and a bit player in this part of John's story, one that began in 1865 as Darwin prophesied at the time:

'It is curious how we are reading same books— we intend to read Leckie & certainly to reread Buckle, which latter I admired greatly before.— I am heartily glad you like Lubbock's book so much.— It made me grieve his taking to Politicks, & though I grieve that he has lost his Election, yet I suppose now that he is once bitten he will never give up Politicks, & Science is done for. Many men can make fair M.P.s, & how few can work in Science like him.—' [226]

With John's journey into politics beginning with two election defeats in 1865 and 1868, both in West Kent,[227] success finally came in a different constituency, the Kentish town of Maidstone, in 1870 where he was also re-elected with a margin of sixty-six votes in 1874. He lost the seat in 1880 to the Conservative candidate, Alexander Henry Ross, but within weeks was back in the House of Commons as MP for the University of London.

The 1870s saw the beginning of a new chapter in John's life in which the focus of his attention turned gradually away from prehistoric archaeology towards applying his ideas on progress and evolution in the political arena. His stint as President of the Royal Anthropological Institute during 1871-3 was his last presidential role within an archaeological or ethnographic organisation until he was elected President of the Society of Antiquaries in 1904. Between 1874 and 1913, John published only a few new articles and one new book (Lubbock, 1911) on archaeological and ethnographic subjects. The further editions of *Pre-historic Times* and *Origin of Civilisation* printed during this period were driven more by the fact that publishers had run out of copies rather than any strong desire by John to present new ideas (Owen, 1999; Owen, 2000, pp.257-8).

His collecting activities changed dramatically during the early 1870s and reflect how John redirected his attention and energies in to interests other than archaeology and ethnography. John was still actively collecting material in 1870-2, but there is a sharp drop in the rate of entries recorded in his catalogue from 1873 onwards.

On 12 September 1872, he went on an expedition to Constantinople and 'Asia Minor' with seasoned travelling companions including the Scottish Liberal MP Mountstuart Grant Duff and his wife, Anna. The party travelled to Vienna and Budapest by train where they embarked on a boat of the First Danube Steamer Company, the K.V. *England*, to take them on the first part of their journey into the east of Europe. The scenery through which the River Danube meandered was breathtaking, cutting through the foothills of the Carpathian Mountains, past the historic cities of Bratislava, Budapest and

Bucharest to the river delta on the Black Sea and the port city of Izmail. Only five years earlier, Johann Strauss had been inspired by the natural beauty and elegance of this majestic river to write the Blue Danube Waltz. The party transferred onto another boat for the final leg of their journey across the Black Sea to Constantinople.[228]

Arriving in Constantinople, John would have discovered a city of amazing cultural diversity unlike anything he had ever seen before. For over 400 years this city had been the cosmopolitan capital of the Ottoman Empire and a place where east met west. As he walked through the crowded streets, John would have brushed shoulders with a vibrant mix of people from all walks of life, including Ottoman Muslims, Greeks, Egyptians, Armenians, French and British. At the Grand Bazaar, built by the Ottomans in 1455, John and his friends would have conversed with merchants sat on wooden divans in front of their stalls and been drawn in by the rich aromas of Turkish coffee being carefully prepared to an ancient recipe.

However, Constantinople was not their final destination. Their real ambition in travelling so far from home was to visit a small place in 'Asia Minor' called Hissarlik, believed to be the site of Troy from Greek legend. From Constantinople they travelled across the Sea of Marmara and through the narrow strip of water separating Europe from Asia, the Dardanelles, to an Ottoman fortress settlement called Çanakkale. This was volatile territory because of its strategic importance for Europe, Russia and the Ottoman Empire. Only twenty years before, it had been a British staging post during the Crimean War and forty years later it was to become a theatre of conflict yet again during the infamous Gallipoli Campaign of 1915-6. Before leaving Constantinople, the party had met the charismatic Grand Vizier, Midhat Pasha, who gave them safe passage recommendations to all the governors in 'Asia Minor'.

Travelling the final few kilometres of their journey overland, John and his companions arrived at their destination in the autumn of 1872. They stayed with a British government official, Edgar Whitaker, but spent time at the farm of Frederick Calvert and his family at Açka Koy called 'Chiflik'. Calvert and his two brothers, James and Frank, were involved in British consular affairs overseas and he had bought the 2,000 acre piece of land in 1847. The plot purchased included part of Mount Hissarlik, an area in which archaeological excavations began due to the Calverts' intrigue fuelled by the long-standing rumour that Hissarlik was the location of Troy. In 1871, Calvert had entered into partnership with Dr Heinrich Schliemann, a charismatic German businessman and amateur archaeologist who had permission to excavate the Turkish government owned part of the site. Schliemann was a self-made man who had earned his fortune during the 1850s speculating in Californian gold dust and supplying ammunitions to the Russian government during the Crimean War. By the age of forty-one, he was wealthy enough to retire and pursue his real ambition of international travel and the quest to discover the mythical city of Troy. Schliemann was a showman, and his personal style and approach to archaeological technique was very different to that of Calvert. Schliemann went on to take all the credit for the subsequent discoveries that made Troy and Hissarlik famous, and his association with Calvert ultimately fell apart.

John visited Hissarlik just as Schliemann and Calvert were at the start of their relationship. The two were already arguing about excavation methods. Whilst he was there, John 'opened the so-called tumulus of Hector at Bunarbashi' and climbed Mount Ida. The group then travelled to the ancient Greek cities of Smyrna and Ephesus, and the Persian City of Sardis. On 12 October, John returned to Hissarlik where he picked up a 'good quartz hammer' that was later accessioned into his collection as number '1032'. Frank Calvert gave him a similar specimen, number '1033'. John also collected various fragments of pottery during his excavations of 'Hector's Tumulus'.[229] The governor of Beiramitch presented him with a weaver's weight taken from a weaving machine.[230] On Friday 25 October, they started for home by way of the Aegean and Mediterranean Sea. Grant Duff and John bathed from the steps of the steamer, and as they sailed past Sicily one morning they woke to a beautiful sunrise and the awe-inspiring sight of the Mount Etna and Stromboli volcanoes. On Wednesday 30 October, they arrived at Marseilles to pick up the train that took them back home to London and John's family.[231]

This trip sparked a love affair for John with the Near East and North Africa. However, it was not accompanied by a renewed interest in archaeological collecting as his activities in this arena started to decline. In 1873, he travelled to Egypt with Grant Duff who was becoming a regular travelling companion. They saw the Sphinx and the Pyramids, walked in the desert and witnessed an eclipse of the moon, visited the temple of Karnac by moonlight and again the following day when he discovered a 'stone hammer in the open court of the great temple'. They visited Aswan, Luxor and the Tomb of the Kings, where John then walked up into the hills by himself and found a number of stone flakes. On Thursday 20 November, they went to see the Colossi of pharaoh Amenhotep III near Thebes and afterwards:

'Then I climbed by myself to the highest point of the hill & found immense numbers of flakes, a few nuclei, scrapers & c.'[232]

They returned to Cairo where they met a 'fine old gentleman', Hikelyan Beg, before heading for home via Marseilles. John's mother had died in February that year, and on his return to High Elms emotions were bound to be a mixture of delight at seeing his family again, excitement about the travels just made, and renewed memories of such a painful loss.

'My mother died. I am thankful to say she did not suffer at the last. If ever there was an angel on Earth she was one. I do not believe she ever said an angry or unkind word to me in my life. Oh dear me! We buried her in our vault at Down. Robert & Mary, Henry, Beaumont, Fred & Catherine, Monty & Sara, Alfred, Edgar & Gerty. Grey he came. Mr. Robinson did the service. I went the last thing at night & sat some time with her. No one ever had a better or kinder mother.'[233]

John recorded just one new acquisition from his Egypt trip in the catalogue of his collection, an African necklace from Aswan.[234] Despite John finding several stone flakes during this trip, there is no reference to this material in his catalogue. It appears that he donated the flakes he found directly on his return to the Christy Collection at the British Museum.[235]

In 1876, John visited Europe twice with his daughter, Amy, and the Grant Duffs. On their spring trip to the small island of Spezzia off the coast of Italy, John and Amy called on Edward Lear at San Remo en route and stopped off in Paris. At some point they also visited Dijon where John bought a stone axe,[236] and Rome where he bought a Japanese anklet.[237] In September, John went on a small tour of Brittany with Amy and the Grant Duffs. They visited a great number of prehistoric stone monument sites, and on Saturday 7 October called on the Comte de Limur:

'Who overwhelmed me with Civilities & gave me several things including a crystal of staurolite, & a piece of [meteorite] which fell at [Cleynerée] at 10.5 am on the 22 May 1869.'[238]

This gift must also have included prehistoric artefacts from Brittany and the Laugerie rock shelter site in the south of France. After his return home on 13 October, John entered the details of flint artefacts and pottery presented by the Comte as items '1120' – '1131' in his catalogue.

On 15 March 1877, Amy married Andrew Henry Mulholland, a member of a famous family working in the Irish linen and cotton business, and son of a Conservative MP. By a cruel twist of fate Andrew died suddenly and unexpectedly less than three months after the marriage, on 2 June. He was on his wedding tour with Amy in Paris while John was in France on a separate trip collecting plants. John had returned to Paris on 3 June where he heard of Andrew's death and found Amy being comforted by her sister, Harriet. On 4 June, they went to the Louvre and John arranged to bring his daughters to Amiens. From here, he and Harriet drove to St Acheul and acquired some flint flakes.[239] They returned home from Paris to London the next day. It is not difficult to imagine that the flint flakes John acquired from St Acheul during that trip would forever hold sad memories of this tragic event in his family's life. When he recorded them in his catalogue he entered the incorrect, but more poignant, date of 2 June[240] - the day Andrew died.

On Saturday 2 October 1877, John and the Grant Duffs left London for a two-month trip through Europe to Algeria. They stopped off to meet people and view the sites in Paris, Geneva, Barcelona, Tarragona, Valencia, Cordoba, Seville and a number of other Spanish towns. He bought gypsy earrings for his wife, Nelly, from the market at Cordoba and a bronze celt and dagger in Seville. On 1 December the party set sail from the walled city and naval port of Cartagena in south-eastern Spain, destined for Oran in Algeria. They crossed the stretch of Mediterranean Sea that separates Europe from Africa and marks yet another transition across cultures. Oran is a city with deep history. It was founded by Moorish Andalusian traders in 903 AD, was captured by the Spanish and then conquered by the Ottoman Empire. Since 1831, it had formed part of French Algeria. The Vice Consul took John and the Grant Duffs on a tour of the city, with John noting of the experience: 'People & costume most interesting'.[241]

They travelled on to the capital city, Algiers, where the Consul-General of Algeria, Robert Lambert Playfair, who was the brother of fellow Liberal MP, Lyon Playfair, met them. They walked through the native quarter which John found to be very intriguing, and with a local botanist, J. Durands, visited the Botanic Gardens and a cave in which some bones and implements had been found. Durands also took John to view a number of houses in the French part of the city, including the 'Campagne Lagier'. On 14 December, John made a successful offer on this property to its owner, Monsieur d'Hincourt.[242] John returned to London on Monday 24 December with Christmas news for the family that he had purchased a property in

French Algiers which he had renamed 'Campagne des Fleurs'.[243] He also brought back the Spanish bronze celt and dagger from Seville, a stone axe and piece of tissue presented to him by M. Gongora and a slingstone from Algiers.[244]

John returned with his eighteen year-old daughter, Conny, and the Grant Duffs in September the following year to take possession of the Campagne. However, Nelly never got to see their Algerian house, or visit the Kasbah, meet the local people and see their customs, or smell the spices in the market place and the flowers in the garden. By now, she was far too fragile to make such a journey. In 1865, Nelly and John had been travelling by train to the BAAS meeting in Birmingham, with Nelly several months pregnant:

'We were coming down here yesterday by the 3.40 train from Paddington, when, about two miles on this side of Banbury, suddenly we began to bump about, and I felt certain at once that we were off the line...The bumping got worse and worse, we were thrown backwards and forwards in the carriage, and though it seemed rather a long while, the only distinct idea I remember was that in a few minutes more we should probably solve many of those questions which interest us so much...At last there was one much worse bump, then a dreadful scrunch, and then everything was still, and it seemed from the contrast almost a supernatural quiet.

'I felt, however, no pain, but my hands and coat were covered with blood. All this happened, as it were, at once. Nelly had fallen into my arms, and though she assured me that she was all right, still, as I found she was bleeding very much, I could not tell how much she might be hurt, or what effect the shaking might have.'[245]

Travelling at sixty miles an hour, a wheel had come off and the engine had plunged into a field dragging the tender with it. No one was killed, but the accident is said to have contributed to Nelly's deteriorating health in the years that followed. Two months after the accident she gave birth to their youngest surviving son, Rolfe.[246]

Nelly died on Sunday 19 October 1879 at the age of forty-three, leaving John a widower with six children – three boys and three girls, the youngest was fourteen.

'In the middle of the night I was called up and found Nelly in a sort of faint from which she never rallied, and she passed away so quietly

that we did not know the exact moment. It is a comfort to think that
I have been with her constantly during the last month.'

(Hutchinson, 1914, p.166)

After burying his wife, John left High Elms with two of his daughters, Conny
and Gerty, for North Africa. On 16 November, he 'picks up' a Palaeolithic
stone implement near the town of Kolea, 17 miles inland from Algiers. This
piece of stone became item '1168' in John's collection[247] and is now housed
at the British Museum in London.[248]

The 1870s proved to be a tough decade in John's personal life, and
perhaps this contributed to the fact that he no longer collected as many
prehistoric stone implements and ethnographic artefacts. Although he still
acquired items after 1872, the personal determination and scientific drive
he demonstrated in the decade before was absent. From 1881 onwards, he
acquired material no more than five times in any one year. Occasionally he
bought coins, but acquisitions were primarily gifts to him by others,
including Lord Derby. He made little effort himself after 1881 to develop
his collection, perhaps with one exception. He kept in touch with Heinrich
Schliemann and his excavations after his visit in 1872. On 9 October 1886,
John set off on the long journey to Athens where he met with Schliemann
on a number of occasions. Schliemann advised John on buying 'ancient
pottery' while John also bought coins from Professor Roussopoulos. On
Thursday 11 November, he said goodbye to the Schliemanns 'who have been
most kind. S. gave me…2 spindle whorls from Troy'. John's catalogue
recorded three acquisitions from this trip:

'1136 Obsidian nuclei, pres. by Prof. Roussopoulos fr. Milos Nov 1886
1140 Spindle whorl from 2nd City at Hissarlik Pr. by Schliemann
Nov 1886
1141 Small axe. Thessaly Pres. by Pr. Roussopoulos 1886.'[249]

By the late 1870s, John had no need to collect so fervently. He and his
Darwinist companions in the X Club and Lubbock-Evans network had
presented the facts and made their case. They were no longer banging on the
door shouting to be heard, their views on the separation of science and religion
and on human antiquity were being listened to and accepted. They had taken
over the scientific asylum and were becoming the establishment. The role of
his collection began to change and reflect this shift, becoming more of a
museum piece telling a story rather than a device to persuade his colleagues.

We have already seen in Chapter 4 how the Darwinists became Presidents of the Royal Society and the Anthropological Institute. On 14 March 1878, John was elected a Trustee of the British Museum, where Evans later joined him in 1885. Evans was elected as President of the Society of Antiquaries and in this role sat as an ex-officio British Museum Trustee until 1892 when he stood down as President but was elected a Trustee in his own right. John was elected President of the Golden Jubilee meeting of the BAAS in 1881. Held in York, where the first meeting had convened fifty years before, John gave his presidential address about the progress of science from 1830-1880, 'a very large bit of work'.[250] Twenty years after the Oxford BAAS meeting where Huxley, Hooker and John had been the outsiders defending Darwin,[251] John now stood up as President and had the undoubted pleasure of publicly crowning his friend, mentor and master as a king of science:

'I have read with pleasure your Address. You have piled honours high on my head.'[252]

From 1881-6, John held the office of President of the Linnean Society. He was invited to open provincial museums and give prizes. He also received honorary degrees from the Universities of Cambridge, Oxford, Edinburgh, Dublin and Würzburg. In 1886, he was also invited to give the annual Sir Robert Rede lecture at the University of Cambridge. Huxley and Tyndall were the only other X Club members to receive such an invite (in 1883 and 1865 respectively). Interestingly, the subject John chose for his Rede Lecture was botanical not archaeological, *On the forms of seedlings and the causes to which they are due.*

Following his publication of *Pre-historic Times* and *On the Origin of Civilization* John's scientific efforts reverted back to his original passion – evolution and selection in the natural world. Although new editions of both books were published after 1870, there is little change in their content compared to the significant amendments between the first and second editions of *Pre-historic Times* during the 1860s. During the 1880s John still collected material whilst he travelled across Britain and Europe, but his diaries began to make reference to plants and insects rather than archaeological collecting.[253] He was frequently asked to give lectures at local societies and he developed a repertoire of talks that focused on ants, seeds, prehistoric archaeology and 'savages'. John continued to write and publish scientific works but the subjects covered shifted significantly. In the 1870s his original published work described his research into silverfish (Thysanura), the springtail organisms (Collembola) found on decaying leaf litter, and his studies of British wild flowers and their relationship to insects. *On the Origin and Metamorphoses of Insects* was also published as part of the Macmillan Nature series in 1874. These themes continued into the 1880s and 1890s with publications on *Flowers, Fruits and Leaves* (1886), *Ants, Bees and Wasps* (1882), *On the Senses, Instincts and Intelligence of Animals, with Specific Reference to Insects* (1889), *Contribution to our Knowledge of Seedlings* (1892) and *On Buds and Stipules* (1899). He strayed into geology with two publications at the turn of the twentieth century on the *Scenery of Switzerland* (1896) and the *Scenery of England* (1902).

In 1892 he also published a volume entitled *The Beauties of Nature* that encouraged a general readership to look at the world around them with new eyes and find inspiration and beauty in what they saw.

'The world we live in is a fairyland of exquisite beauty, our very existence is a miracle in itself, and yet few of us enjoy as we might, and none as yet appreciate fully, the beauties and wonders which surround us.'

(Lubbock, 1892, p.3)

It was one of four books he published in his later years that deliberately intended to educate and inspire a wider audience about life. The *Pleasures of Life* was written in 1887 and formed a compilation of addresses he had given to young boys and girls at school and college prize-givings. *Use of Life* (1894) and *Peace and Happiness* (1909) were the other works. All of these volumes picked up and expanded upon the ideas of social, moral, technological and economic progress that John had first formulated and expressed in his works on prehistory and ethnography during the 1860s and early 1870s. These were the ideas that also drove his political agenda; for forty years John championed the cause of education and opportunity for working men, women and children across the country.

It was to the political arena that John attentions really turned after he was elected MP in 1870. This was where he now wanted to make a difference and his stated aims upon entering Parliament were:

'To promote the study of science, both in secondary and primary schools, to quicken the repayment of the National Debt, and to secure some additional holidays, and shorten the hours of labour in the shops.'

(Hutchinson, 1914, p.110)

He did not drag his feet. One year into his term of office he managed to pass the very popular Bank Holiday Act. The third reading of the Bill was passed on 15 May 1871 and shortly afterwards legislation existed that identified Easter Monday, Whit Monday, the first Monday in August and Boxing Day as Bank Holidays. On these days, banks would be closed to business and nobody could be compelled to do anything that they were not required to do on Christmas Day or Good Friday. On Monday 7 August 1871, bank clerks, shop workers, warehouse labourers and office staff took a paid day off work to spend with their families:

'[Sir John Lubbock] has added – substantially added – to the sum of human happiness, and has carried rays of hope and joy into humble households so great as to rank him high as a public benefactor.'[254]

At the age of thirty-seven, John had introduced an idea that gave ordinary people the chance to enjoy the world around them and look beyond the routine of work, eat and sleep. This was seen to be a true sign of civilization if ever there was one, and the people loved him for it. Despite the long queues and crowds trying to board steamer boats and trains for the British seaside, the press wanted to call the first Monday in August, 'St Lubbock's Day'. Others wanted to present him with a testimonial in the form of a silver type of Darwin's ape (Hutchinson, 1914, pp.123-4).

John continued to lead or support attempts to introduce legislation that gave working people more time to spend with their families, visit places and educate themselves about the world. He played a significant role in passing the Shop Hours Regulation Act, which limited the number of hours worked by a person under the age of eighteen to seventy-four hours per week. It took two years of hard work from introduction of the first Bill in 1884 before it was enacted into law. His attempts to secure early closing of shops (which could be legally opened all night and day with shop assistants compelled to work) took even longer. An Early Closing Movement had been established in 1842 and John became a leader of this movement after 1870. He first presented a Bill to Parliament about the early closing of shops in 1873 but failed to get a second reading. In 1888 he raised a Bill calling for shops to

105

close at 8pm and 10pm on Saturdays but again it was rejected. It was not until 1904 that the Shop Hours Act was passed.

In the House of Commons and elsewhere, John also sought to champion the cause of science and education. In March 1871 he spoke on elementary education, stressing the importance of including the subjects of history, geography, English and elementary science in the curriculum. This proved too radical a suggestion at this early stage in his political career. Over the next few years, John was involved in supporting the passage of a series of educational Bills into law. The successful 1876 Elementary Education Bill required parents to ensure children aged between five and thirteen received a basic education in reading, writing and arithmetic. Parents had to pay a few pence each week in fees, and boards of poor law guardians were able to provide financial assistance if required. In 1880, a new Education Act made school attendance compulsory for all children up to the age of ten. Elementary education was made free in the Free Education Act of 1891 and in 1892 the Public Libraries Act was passed that inspired the immediate growth of local free libraries across the country (Patton, 2007, p.7).

In 1872 John was elected to the position of Vice Chancellor at the University of London. He had objected, given that he did not have a degree and felt himself to be under qualified, but members of the Senate overruled his concerns. He was re-elected to this post until 1880 when he became MP for the University of London. During his tenure, the University became the first in the country to accept women for graduation in all faculties.[255] As a Trustee at the British Museum, he proposed that the Museum open in the evening for working people to visit, a proposal that was agreed.[256]

Until 1885, John had an impressive political record and was a rising star within the Liberal Party. He had been instrumental in passing important finance and banking legislation and was an ally of the Liberal Party leader, William Gladstone. Gladstone was an amateur classical historian and shared with John an interest in archaeology and Schliemann's excavations at Hissarlik. On 12 March 1873, John recorded in his diary a conversation that he had with Gladstone about politics and archaeology.[257] On Friday 25 June 1875, John breakfasted with Gladstone so they could meet with Schliemann.[258] The German entrepreneur was a clever salesman and knew how to gain public kudos for his work, overcoming any professional disquiet about the destructive excavation methods he used in his search for Troy. His connection with the popular and charismatic figure of Gladstone was a valuable one, and he persuaded this enigmatic figure on more than one occasion to publicly express support for the importance of his work. On Saturday 10 March 1877, Gladstone came to High Elms for the weekend with Miss Gladstone, Huxley and Lyon Playfair:[259]

'We walked up to see Mr. Darwin. Gladstone has made himself very pleasant. He is full of Turkish iniquities.'

At this time, Gladstone had stepped back from politics after being bitterly defeated at the 1874 election by his political and personal archrival, Benjamin Disraeli. In 1880, the Liberals won back power and Gladstone was appointed Prime Minister for the second time. Lord Derby, who defected from the Conservative Party in the same year, served as Colonial Secretary in Gladstone's government. On 10 October 1883, John travelled to Liverpool and visited his political colleague, Lord Derby, at Knowsley. On the next day, John visited the 13th Earl of Derby's collection at Liverpool Museum and took a walk with another Liverpool son, Gladstone himself, during which they discussed Homeric myth and politics.[260]

Perhaps even during this walk, future tensions between John and Gladstone might have begun to surface. For these were turbulent times in British politics and for the Liberal Party in particular. Gladstone's second Parliament waned in popularity over its five-year term, not least because of decisions made regarding foreign policy. The event that ultimately inspired its downfall took place thousands of miles away in an African city; where the Blue and White Nile meet and flow northwards as the River Nile through Sudan and Egypt to the Mediterranean Sea. A Victorian national hero of many years standing was slain by Sudanese rebels led by Muhammed Ahmad al-Mahdi. His name was General Gordon, the city was Khartoum and the date was 26 January 1885. Gordon had been sent to support the evacuation of Egyptian soldiers from the city and instead lost his life. Gladstone and his government were blamed for not doing enough to ensure the success of this operation.

On 8 June 1885, John returned from a trip to Switzerland to find that the Government had resigned.[261] Gladstone lost the General Election and Lord Salisbury established a minority Conservative government. Determined to win back power, Gladstone made the decision to throw his lot in with the Irish Nationalists who now held the balance of power. He succeeded in defeating Salisbury's government and became Prime Minister, but the price paid was more than John and Lord Derby were prepared to accept. John had already had some difference in opinion with Gladstone regarding foreign policy in Egypt and questions of electoral reform. John was an active supporter of proportional representation and was elected the first President of the Proportional Representation Society when it was founded in 1884. However, the political dynamite that finally split John and Gladstone also sent permanent shockwaves through the Liberal Party.

Gladstone formed his new government in November 1885, and there is a theory that Gladstone might have been considering John as a candidate for Chancellor of the Exchequer (Hutchinson, 1914, p.222). However, John would have been unlikely to accept because of the deal Gladstone had made with the Irish Nationalists. To gain power from the Conservatives, he had committed to introducing a Home Rule Bill that would have granted self-government to Ireland. John was one of many Liberal MPs who fundamentally opposed Irish Home Rule, fearing it would result in an independent Ireland and dissolution of the United Kingdom. Led by Lord Hartington and Joseph Chamberlain, this group of MPs, including John and Lord Derby, formed a political alliance with the Conservative Party to oppose Gladstone. On 8 June 1886, Gladstone's Bill was defeated by thirty votes in the House of Commons. The Liberal Government fell, and John became a member of the break away Liberal Unionist Party.

Throughout this period, the X Club and Darwin still held a special place in John's life but the meaning of these relationships changed as the time marched on towards and beyond the end of the 1870s. For example, on Sunday 4 January 1873 and Sunday 25 February 1874, the Spottiswoodes visited High Elms and popped over to see Darwin.[262] On 8 April 1874, John took the children to Alum Bay on the Isle of Wight. Nelly was too ill to make the trip but Tyndall came and went climbing. Slightly embarrassing for a climber of his pedigree, he got stuck and had to be rescued with a rope. John noted in his diary that 'he had a narrow escape'.[263] On 3 October 1874, Spencer and Tyndall came to High Elms for Sunday with Max Muller and Grant Duff, and later that month Spottiswoode and Evans came down to spend the weekend. On Monday 7 December, Huxley also visited High Elms for the day.[264] On 25 November 1876, the Hookers came for Sunday at High Elms, and on 2 December the same year John walked with Spottiswoode to see Darwin.[265] It was Nelly's birthday on 15 February 1877 but he dined with Huxley, Hooker and Spencer at the Athenaeum.[266] Throughout the 1870s and into the 1880s, John continued to attend X Club dinners and referred to a number of them in his diary.[267]

The picture that emerges is one of friendship and belonging compared to the scientific zeal that brought these compatriots and their families together in the 1860s. On 28 April 1883, four years after Nelly died, John described High Elms with the words 'always makes me sad to go there, the place is so full of memories sad ones', and it is clear from his diaries that he found solace in spending time with the friends he made in the decades before. On 27 May John called in at the Walsinghams where Pitt Rivers, Spottiswoode and Mrs Busk were. On Sunday 10 June he called on the Busks and two days later met Huxley in Cambridge when he was staying with the Vice-Chancellor. On Saturday 21 July, John went to Hindhead with Mrs Tyndall where he met up with Tyndall and the Hookers. The next day he accompanied Hooker on a flower-spotting walk. On 25 July he dined with the Huxleys and a month later stayed with the Hookers at Sunningdale. However, we also know that later this year, relationships between John and at least some members of the X Club became strained for a while when Huxley was elected President of the Royal Society on Spottiswoode's death.[268]

It is through these long-standing friendships that John found new personal happiness during 1883 after a chance meeting with the twenty-one year-old daughter of Pitt Rivers, Alice. On Saturday 17 November 1883, John and his daughter Gerty made their first ever trip to the Pitt Rivers' family home at Rushmore, near Salisbury. Alice was late coming down to dinner and had been reprimanded by her mother. John lent her a sympathetic ear and in his diary entry for 18 November noted 'I had a good deal of talk with Alice Fox Pitt'.[269] On Thursday 6 December, he commented that he walked with Alice at High Elms and they talked about religion. That evening he went up to the X Club and the Linnean Society meeting. Five days later John recorded the fact that Alice left High Elms and annotated this entry with the underlined words 'got quite fond of her'. Four months later he travelled to Rushmore with his youngest son and daughter, Rolfe and Gerty, to ask Alice to marry him:

'Delightful walk with Alice to the Museum...I then told her how much I loved her, but that the difference of age made me feel that it was not fair to ask her to marry me. However, she was a dear & accepted me, & after a short walk in the woods we came in.'[270]

Twenty-nine years separated them in age but John was a wealthy, energetic, charismatic and intelligent man. The couple married on Saturday 17 May 1884 with the Hookers amongst the guest list. They departed for their honeymoon to Paris, Switzerland and Italy on 13 June where they spent some time collecting and recording flowers. John had already introduced Alice to

Darwin's *On the Origin of Species* a few days before their wedding[271] and she began to take on an assistant role in a similar way to Nelly all those years before. John's children, John, Norman and Gerty, nicknamed Alice 'the Ancient Monument'![272]

Less than two months into their marriage, Alice and John were arranging an educational open day for local people to look at his archaeology and ethnography collection, and observe the behavioural experiments he was undertaking with his ants and a dog called Van.

'In the afternoon [Saturday 5th July, 1884] 60 members of the Sidcup and Crays Natural History Society came. We gave them tea & showed them the Museum, Ants & Van.'[273]

Here we immediately see the influence of Alice on John's scientific interests (Plate VI). This was exactly the sort of wider educational work Alice's father, Pitt Rivers, believed was so important and undertook on a vast scale with his now large collection.[274] In 1891, John noted that five local groups came to visit High Elms and look at the 'museum'. On 12 January and 19 October local people from Downe and Farnborough visited. On 16 November, the Bromley Literary Institute, Bromley Natural History Society, and the Working Men of Downe and Farnborough came. Similarly, on 24 October 1892 about 200 people from Downe and Farnborough came for an evening soirée to view the collections.[275] The Bromley and District Teachers' Association visited on 3 September 1898,[276] and even in September 1912 – a few months before John died –natural historians from across the country descended on High Elms:

'The Selbourne Society came for an outing, about 100. We gave them tea, I took them round the ponds & put out some things on the billiard table. We rather dreaded it, but having put them off twice, did not like to do so again. They seemed interested. Poor Alice was tired.'[277]

Very little is known about how John stored or displayed his collection before 1890; family lore suggests that items and the paintings commissioned from Ernest Griset[278] were displayed in the hall, library and billiard room.[279] We know that in April 1872 Napoleon III, who was staying in exile in Chislehurst at the time, visited John and viewed his collection of antiquities, remarking that he wished to imitate his example by creating a similar museum at St Germain on his return to France.[280] We also know that on the day of his mother's funeral, Tuesday 18 February 1873, the museum was hung with white and evergreens.[281] In August 1890, John, Alice and their little daughter Ursula travelled to Switzerland for a vacation during which they collected plants, made some geological observations for a book John was preparing, and met up with his old friend, Tyndall. They returned home hoping that the workers would have finished their work redecorating the hall, only to find the workmen still there, 'as usual'. John spent the next two weeks arranging the display of books, flint implements and 'savage things' in the Hall with the help of Mr Oldland from the British Museum. Oldland worked for the curator, Franks, and was probably fixing John's material with hooks and wire loops at Franks' suggestion so that the specimens could be taken down for closer examination if required.[282] No photographs or illustrations exist of how items were laid out, but the close connection with the British Museum suggests John may have adopted some of the display techniques used by Franks. After all, as a Trustee of the British Museum, John had attended the opening ceremony for Frank's new gallery in April 1886, and would have known it well.

In 1900, John had the collection re-hung in the library and hall of High Elms, again with Mr Oldland's assistance:

'Mr. Crook & Mr. Oldland have been finishing up the library & the savage Implements in the Hall.'[283]

This time we do know more about the display layout because John described it in detail at the back of volume two of his collection catalogue. In the Library, the wall space was divided into eight alphabetical zones (A-H), two numbered zones (1-2) and six descriptive zones ('Over door', 'Over fireplace', 'End', 'Over window', 'On cross beam', 'Over window'). The collection catalogue also gives a list of the items displayed. There does not appear to be a clear logical pattern of display with several zones containing a mix of cultures and objects:

'2 Clubs Guiana
Arrows S. America
Pillow S. Africa (double)
Pillow E. C. Africa
Pottery Algeria
Paddle British Guiana
Bows S. America'.[284]

The Hall contained eight groupings of material, mostly throwing sticks, paddles, clubs and shields from Australia, the Pacific Islands and Africa.[285]

The stones and bones of John's collection tell us a story of someone who, by 1880, had moved on to new things. Their meaning and purpose had transformed from scientific research tool to semi-public educational resource. From the exploits of youth to the personal nostalgia that comes with age and experience:

> 'Alice & I joined the Evans' & Prestwich at Dunton Green & went on by Westerham to Limpsfield to Mr. Bells to walk on the ground with him where he has been finding flint implements. It was a dull grey day, but I was very pleased to have another day in the field with Evans & Prestwich & it carried me back to old days on the Somme & elsewhere, 25 years ago!...We found a few flakes, but nothing much.'[286]

On 12 September 1885, Leonard Lyell presented John with a stone axe from St Acheul collected by his father, Charles, over twenty years before: another souvenir to add to his collection to remind him of the heady days of Darwin's brothers in arms. John and Alice were staying with the Lyells for a few days whilst on a tour of Scotland.

In these later years, the artefacts in his collection were wrapped up in international stories of discovery, as well as connected to the memories of people once known and loved. The artefacts from prehistoric burial mounds in Denmark picked up in person by John and his first wife, Nelly, in 1863; the flint flakes collected from St Acheul two days after Amy's first husband died; the precious stone implements and ethnographic artefacts given to John by his X Club colleagues, now passing on; the bone harpoon from Tierra del Fuego received as a gift from his greatest mentor, Darwin, whose 'life ended at 4 o'clock in the afternoon, Wednesday 19 April, 1882' (Desmond & Moore, 1991, p.663).

Chapter 8

The Stone Circle

H er hand wavered for a moment over the chessboard. Should she play defensively by moving her noble white Knight or go on the counterattack with her impatient Queen? Her opponent was on form tonight and taking no prisoners, yet her decision was made and the Queen moved three squares to the right. She looked up at her friend sitting opposite and smiled as she reached for her small glass of Madeira wine from which she took a large sip. The two amiable combatants were sat in a room at the Ailesbury Arms Hotel in the Wiltshire market town of Marlborough. Anna Grant Duff and John were staying here for a few days with her husband, Mountstuart, and some of their friends.[287] They had come on a tour of the prehistoric sites in this part of the world, close to home. In particular, they planned to visit the largest group of stones John ever acquired for his collection.

Nine months before in 1871, John and his daughters (Harriet, Amy and Conny) had caught a train from London Paddington to Uffington station in Wiltshire. From there they picked up a carriage to take them to the Iron Age hill fort at Uffington Castle and view the spectacular White Horse carving close by. Almost 400 feet long and cut into the hillside by our prehistoric ancestors 3,000 years ago, it rarely fails to take visitors' breath away, even today. Imagine what it must have meant to John and his family at a time when 'prehistory' was an idea still so new and little understood. The small family party also visited Stonehenge, an awe-inspiring prehistoric stone circle in the middle of a bleak landscape on Salisbury Plain; and the mysterious Silbury Hill, the largest manmade earthen mound in Europe constructed from chalk.

However, it was the uniquely special and quirky group of stones scattered around the small village of Avebury, within sight of Silbury Hill that truly caught John's imagination:

'There is a village amongst the Wiltshire Downs lying in a hollow below broad green pastures and chalky hills. It has but one long street and a few straggling cottages and grey farmhouses amongst gardens and trees—happy and homelike as an oasis in the desert to the traveller who first looks upon them from the heights; and near it and within it stand smooth stones, giant in size, and deep and mysterious in their meaning, the relics of a heathen worship; and high grassy banks, upon which children play, and along which labourers plod, without a thought of the history pictured before their eyes, mark the precincts of those ancient temples. In the centre of the village is the Rectory...fronting a pleasant garden and green fields, across which was a path leading to a vast mound said to be the work of human hands. Marvellous it is even as the mystic stones that tell of the creed of the generations gone by; and solemn and peaceful are the blue mists that rest upon it in the early morning...I have lingered at the garden gate day after day, gazing upon the old circular hill, and hearing no sound to break the stillness of the air, until I could have fancied that peace—the peace of a world which has never echoed to the sound of a human voice—the peace of the spirits who rest in hope, was lingering amidst that quiet village.'

These words were written by a Victorian traveller, Miss Elizabeth Sewell, inspired by a visit she made to Avebury over twenty years before.[288] They sat well thumbed on the library shelves of the village rectory where the reverend Bryan King sat down for lunch with John and his daughters on 30 June 1871.[289] One can only imagine the conversation around that table. They may have commented on the sheer wealth of old stone monuments to be found in the neighbourhood, and would probably have spoken about the antiquarians who thought Silbury Hill and Stonehenge were Roman in date. John would undoubtedly have discussed his belief in their prehistoric origin.

We do know that John had remarked that he might be interested in buying the two meadows containing the stones and dyke if they ever came up for sale. Bryan King remembered this comment. On 3 October 1871, King wrote to John informing him that a building society had bought the farm owning the meadows, and that they had put one of the meadows on sale for £550. The meadow on the market did not contain most of the stones, but it was an area of approximately six acres and would represent a good annual investment, letting for £15-£16 per year.[290]

So began John's acquisition of a prehistoric stone circle. In contrast to the rest of his collection, these stones could not be held in the hand (Plate VII). The outer circle stones were over three metres tall and many tons in weight. He bought his first part of Avebury at a distance, sat in his offices in the City. Mr Thomas Kemms, the owner of Avebury Manor, acted locally on his behalf, instructed by Bryan King. He conversed with the solicitor and found that the meadow for sale did not include the mound or dyke. John was adamant, he expected to buy the dyke as part of his £550. Kemms met the parties who did hold ownership and persuaded them to give up the meadow in return for other land nearby. Two-thirds of the small holders expected £1 in gratuity, and the largest holder requested that he might rent the land back from John after purchase.[291] In the end, John shelled out £563, 16 shillings and 9 pence on the Avebury Estate during 1871-2.[292] In the days before photographs, all John had to show for his purchase in November 1871 was a scrappy sketch drawn for him by Bryan King.[293]

It was not until Friday 29 March 1872 that John had the satisfaction of taking his friends, the Grant Duffs, and their companions to stand on his Avebury land for the first time.[294] He also had the warm glow of knowing that he had helped to protect this curious but, in his view, important archaeological monument from further destruction. The stone circle at Avebury was a shadow of its former self even when John came along to rescue it. Centuries of stones being reused for gate posts, removed to allow farming, and houses being built in the circle itself, had all taken their toll on a henge monument four times the size of Stonehenge and older in date. John had responded to the call of local archaeologists and scientists fearing that the land would be sold to a developer who would build more houses on it. To them, he was their champion:

'At a Committee meeting of the above Society [Wiltshire Archaeological and Natural History Society] I had the honour of moving that "a special vote of our most cordial thanks be offered to Sir John Lubbock, who with a public spirit above all praise and the true love of Archaeology for which he is notorious, came forward at the right moment to rescue the great Circle at Avebury from the profanation with which it was threatened." I need hardly say that this motion was heartily agreed to.'[295]

On the national stage his contribution was also noticed. Joseph Burtt wrote to John asking him to attend the next meeting of the Royal Archaeological Institute of Great Britain and Ireland to talk further with members about 'what you have so nobly done to save the grand monument of Avebury'.[296] John recorded the event in his diary with the simple sentiment, 'I have bought part of Abury to prevent it being cut up for building...I am very glad to possess part of Abury.'[297]

Darwin's Statesman

When John finally succeeded in becoming an MP in 1870, Darwin's apprentice began to take on the mantle of his Master's statesman. His successful introduction of the Ancient Monuments Protection Act 1882, by which the state recognized the importance of prehistoric sites as evidence requiring protection, was probably one of John's greatest Darwinist achievements. The Act laid the foundations on which all other heritage preservation legislation now rests. Little did he know at the start how difficult a challenge he was undertaking. Passing the Bank Holidays Act was easy in comparison to a mission that took ten years of hard political graft to complete. Not only did he have to win the intellectual argument that these monuments were an essential part of the scientific story of human antiquity in Britain and Western Europe, he also needed to persuade his parliamentary peers that they were worthy of public investment and protection over the private interests of landed gentry. For John there was also one further challenge – the subject of ancient monuments protection was not a priority for either Gladstone's Liberal government or Disraeli's Tory government in the 1870s. He had no choice but to introduce his ambition as a Private Members Bill, a notoriously difficult route with a low success rate. It meant that he and his fellow supporters had to promote their cause, fight for Parliamentary time and sponsor all the work involved themselves.

His active personal involvement in protecting the prehistoric monuments of Wiltshire was at the heart of this wider ambition. He had first visited Stonehenge on 22 June 1862 with Nelly, Busk, Christy and his younger brother, Edgar. From the very first, he was 'delighted' with it. In 1866, he became embroiled in a debate with James Fergusson in the letter pages of London's Athenaeum Club journal. Fergusson believed passionately that Stonehenge and Silbury Hill were Roman in date, however, John argued for their prehistoric origins in the Neolithic and Bronze Age:

> 'He [John Lubbock] does not, either in his book [*Pre-historic Times*] or in his letter, produce a single historical testimony in favour of his views, nor one tangible analogy from any other building of the Bronze Age, nor one fact strictly applicable to the monuments themselves.'[298]
>
> 'Mr. Fergusson regrets that I did not, either in my book or in my letter, "produce a single historical testimony in favour of" my views. I regret this also; but we cannot have historical testimony of a pre-historic fact.'[299]

John was confidentially asked to participate in excavations at Stonehenge by the archaeologist Thomas Wright, explaining 'the first in my mind is yourself',[300] after having been commissioned to secretly form a team of six people. John took part in the excavations at Silbury Hill in 1867. From that date he had a friend and supporter in the reverend Alfred Charles Smith, a resident of Wiltshire and the Secretary of the Wiltshire Archaeological and Natural History Society.

> 'I cannot address you as a stranger having had the pleasure of meeting you at the British Association at Bath and once since in London. I send you by this post a copy of the record of some excavations the Wilts. Arch & Nat Hist Society...made some time since at Avebury.'[301]

These records supported John's prehistoric theories by providing further evidence that Avebury and Silbury predated Roman features in the landscape.

It was Smith who moved the Society's vote of thanks to John for rescuing Avebury in 1872, and it was he who wrote on 19 June 1873 asking John to be their next President with the words 'we want a good Archaeologist to preside over us; and I am sure we cannot find a better than yourself'.[302] John accepted the invitation and travelled down to Wiltshire to give his Presidential Address in the same year that the Society opened its brand new museum in Devizes. John bought Silbury Hill for £500,[303] and two years later took the opportunity to buy more of Avebury:

'Down to Abury to see my new little property. Liked it very much. There are two delightful little meadows, & it joins onto the piece already bought.'[304]

In the following year, John was back to attend the Wiltshire Archaeological Society meeting in August 1876, and to revisit old arguments about the date of Stonehenge:

'At Stonehenge there must have been between 4-5000 people. Parker spoke and made it out post Roman & I was put up to answer him. The audience was very attentive, & I was sorry I had not prepared something to say.'[305]

Despite John's attempts to preserve prehistoric monuments through purchase, he was very conscious that many important archaeological sites were being destroyed on a daily basis. In March 1872, on the same trip that he first saw his bit of Avebury, he also saw evidence of destruction:

'We went to the stone Dolmens which are marked on the OS map. The first is still perfect. The second is falling to pieces. The third, that in Temple bottom, has disappeared altogether, having been removed by Mr Tanner the farmer since my first visit.'[306]

He had known for a long time that something needed to be done on a national scale to prevent this vandalism. As early as February 1863, John had written to Steenstrup in Denmark about his desire to provide legal protection for ancient monuments:

'I am anxious that we should have an inspector of Antiquities as you have in Denmark, many of our most interesting monuments of prehistoric times are perishing or being destroyed for want of some such arrangement. Several of my friends to whom I have spoken on the subject would support an application to Government, but before moving in the matter it would be very desirable to know how the arrangement works in Denmark, & if you could forward me any information on the subject I should be very glad.'[307]

Now that he was a backbencher in the House of Commons, he was determined to do something. John identified the introduction of legislation to protect prehistoric ancient monuments from destruction by farmers, housing developers, canal builders and railway planners, as one of his primary objectives. His crusade began in earnest in the latter months of 1872 as he consulted and lobbied behind the scenes to prepare his first draft Bill. On 22 November, he wrote to fellow MP, Osborne Morgan:

> 'The archaeologists of this Country are very anxious to stop if possible the continual destruction of Prehistoric Monuments, & at last we have agreed on a Bill & Schedule which has met with general support. Mr. Bouverie & Mr. Beresford Hope have agreed to go onto it with me, & I write to ask if you could give us your valuable cooperation. I may add that Sir Roundell Palmer had looked through the Bill & generally approved...Mr. Lowe has seen the Bill but we could not get anything definite from him.
>
> 'I send you the Bill & need not say that we should be most glad to consider any suggestions you may make.'[308]

Mr Morgan joined John, Bouverie, Beresford Hope and Mr Plunket on the Bill and together they presented it for a First Reading in the House of Commons on 7 February 1873. With John and his cosponsors sitting at the table, a Clerk read out the title of the Bill in the House to let other Members know of its existence. This ritual was over in just five minutes, but for the first time in history it was on record that Parliament was being asked to consider introducing legislation to protect British ancient monuments. At the Second Reading on 6 May, John briefly introduced the Bill to cries of support from backbenchers. Mr Bruce also spoke on behalf of the government and agreed to support the Bill as long as provisions that required the spending of public funds on purchase of monuments were removed. Although undoubtedly disappointed, John was unsurprised and accepted these conditions. The Bill was read and committed to on the Friday, but no reference to it appears on the minutes of Friday's House of Commons Sitting, and it disappeared from view.[309]

On 20 March 1874, John re-presented the Ancient Monuments Bill for its First Reading in the Commons. Mr Russell Gurney and Sir William Stirling Maxwell joined Beresford Hope and Morgan as cosponsors. It was scheduled to receive its Second Reading after the Easter break, on Wednesday 15 April 1874. On this occasion, John presented a lengthy argument for need of such legislation, drawing upon detailed evidence of monuments already lost and examples of legislation from other European countries that put Britain to shame. Instead of cheers he was met with fierce opposition from the Tory MPs of Disraeli's new government. Many objections were made including the perceived interference of the state in matters of private property, and whether the purchase of sites by the nation was a good use of public funds. The House divided and the ballot opened for Members to vote in favour of the Bill proceeding to the next stage with 'Aye', or not with 'Noe'. The latter won the day, with 147 votes to 94, and the Bill was thrown out.[310] 'The opposition was rather a surprise',[311] leaving John and his supporters angry that no advance notice had been given about these concerns, as was usual practice.

The final nail in the coffin had probably been the objection presented by the Tory Chancellor of the Exchequer, Sir Stafford Northcote. He suggested that John's Private Members Bill was unconstitutional in that it involved the allocation of public funds to new purposes. Only government sponsored Bills could introduce such changes and he proposed that if there was a need for such a Bill the government should be asked to consider it.[312] John's attempts to reintroduce in 1875 fell at the same hurdle but not without causing some drama and excitement first. A very similar Bill to that presented in 1874 was given its First Reading on Monday 8 February 1875. John then led a much better prepared deputation for its Second Reading on Wednesday 14 April. The questions posed and comments made by other Members of the House were numerous and represented both supporters and opposition. This time when the House divided and Members voted it was John's turn to feel 'excited' and proud. They beat the Government by twenty-two votes and the Bill was given its Second Reading.[313] However, their joy was short-lived. On 17 June, John was forced to withdraw the Bill when the Chancellor of the Exchequer repeated that he could not support its passing for reasons already described in 1874. However, he would be happy to recommend to his government colleagues that the matter of ancient monuments protection is looked at in the autumn.[314]

No progress was made and John was forced to present yet another unsuccessful Bill in 1876. He and his friends were not prepared to give up, and their tenacity began to pay off in 1877 when they finally managed to pass their Bill through the Second Reading to the Committee Stage. At this point, a smaller group of Members sit down and work through the detail of the Bill, discussing and voting on possible amendments. Committees usually only last a number of weeks and on this occasion the Committee sat down for their first meeting on Monday 25 June 1877. John was elected the Chair and by Monday 9 July, he reported back to the House that the Bill had passed Committee. However, it was a case of so near and yet so far – on 23 July the Commons ruled that there was no more time to consider Private Members Bills before the end of Session and the summer break. John and his ancient monuments had missed out again!

In 1878 and 1879 his attempts failed even to get out of the starting blocks, and any lesser mortal would probably have given up. However, for John this was a mission that lay at the heart of his Darwinist scientific beliefs; he undoubtedly had the support of Darwin, the X Club and the Lubbock-Evans Network to help him succeed. In the dying days of Disraeli's weakened government in 1880 and immediately after the Opening of Parliament on 5 February, John gave notice of the Bill and got it read a second time. He was in no mood for being played this time and so when questioned about whether the lateness of the hour (2.30am) was appropriate, he quite simply replied:

'He would have been very reluctant to press the Bill at that hour of the night, had it not been for the manner in which it had been treated on previous occasions. If his hon. Friends opposite really desired to consider the Bill, he would be happy to give them time to do so; but he feared that they had made up their minds. The Bill was really the same as that upon which the House had often expressed an opinion, and he thought, therefore, that no object would be gained by delay.'[315]

Perhaps it was the time of night that made him cut right to the point, or perhaps his wife's premature death less than three months before reminded him about the importance of seizing the moment. On Tuesday 24 February 1880, he was able to record in his diary:

'At last I have got my Ancient Monuments Bill through after nine years work. How pleased Nelly would have been. Lord Stanhope is going to take it in the Lords.'[316]

It wasn't until 18 August 1882 and under a Gladstone government that the Ancient Monuments Protection Act finally became law.

Unsurprisingly, the final piece of legislation did not achieve as much as John and his supporters had set out to do in 1872 as compromizes had been made on the way to create a more palatable Bill that had a chance of passing into law. In the end, the Ancient Monuments Protection Act invested a limited degree of power in the newly established Commissioners for Works responsible for ensuring the protection of selected prehistoric monuments in England, Wales and Scotland. These monuments were listed on a schedule attached to the Act that had been prepared by John in 1878 while working with the Society of Antiquaries. Owners could voluntarily hand over control of their monument to the Commissioners whilst retaining ownership; in return the Commissioners would accept the duty of maintaining it. It became a punishable offense to damage any of the listed monuments, whether they were in the care of the state or not. The state gained the power to buy any of these monuments that might be offered for sale, or to receive them by gift or bequest. John and his colleagues were only able to succeed because they redirected the debate from one about private ownership to the notion that no one had the right to destroy nationally important monuments like Stonehenge (Chippindale, 1990).

The Act also introduced an 'Inspector of Ancient Monuments' whose role was to monitor the condition and protection of the monuments on the list. In a letter to Pitt Rivers, dated 25 October 1882, John wrote:

> 'George Lefevre is going to talk to me about the Inspectorship of Ancient Monuments & I should much like to know your views as to the Salary, Conditions, & duties etc. which would be suitable. I suppose it must be a fixed salary & expenses? Who do you think would be the best man?
>
> 'You know the monuments as well as anyone, but I presume you would not think of it.'[317]

John's future father-in-law, Pitt Rivers, agreed to take on the role and was appointed by the Treasury on 25 November.

Nelly was not the only intimate person in John's life with whom he could not share the excitement of this crowning achievement. On Sunday 19 March 1882, John went to visit Darwin at Down. It is hard to imagine that he did not update his master on all that was happening in Parliament and the halls of London scientific power. However, on this trip John would have found a fragile shell of the man he once knew, racked with pain and awaiting his own death. Twelve days before, Darwin had been shuffling along the Sandwalk as usual when he had a seizure and collapsed. His doctor confirmed angina and that there was little that could be done except manage his pain.

For a while now, Darwin had opened up and discussed his thoughts on

God and religion that he had kept so firmly private during the debates of years gone by. It is very possible that John and his master spoke about death, belief and the idea of an afterlife during their meeting on 19 March. John was still a believer in God as well as a disciple of evolution and natural selection. Darwin would have been content now to share how he had given up his Christian faith at forty years of age; how the tragic deaths of his father in 1848 and his precious daughter, Annie, in 1850 had sealed his spiritual fate (Desmond & Moore, 1991). Later, John recorded in his diary that this was 'the last time I saw him'.[318] Exactly one month to the day, the great man died in his wife's arms.

'For thirty years he has been very good to me, and a talk with him was as good as sea air.'[319]

Their relationship had soured slightly a few years previously over negotiations regarding Darwin's purchase of land from John that incorporated the Sandwalk. However, there was no doubting their strong and loyal connection to the last. Before he died, Darwin expected to be buried alongside his family in the small and peaceful village churchyard at Downe. However, his death had barely been announced when moves were being made by the X Club and the press to have him buried at Westminster Abbey. Knowing Downe and Darwin well, John's natural instinct was to support a local burial with friends and neighbours. He also recognized his wider duty and the call to arms from Huxley, Spottiswoode and others. On 21 April 1882, John wrote a letter to the Dean of Westminster on House of Commons headed notepaper and arranged for it to be signed by as many colleagues in Parliament as he could muster:

'We venture to suggest that it would be acceptable to a very large number of our countrymen of all classes & opinions that our illustrious countryman, Mr. Darwin, should be buried in Westminster Abbey.'[320]

Signatories included the geologist and Liberal politician Nevil Maskelyne, as well as Sir George Otto Trevelyan, Lyon Playfair, the Earl of Derby, Henry Campbell Bannerman and Thomas Brassey. The X Clubbers lobbied in various corridors of power and the liberal Dean of Westminster agreed to the collective request. Spottiswoode, John and others persuaded Darwin's family to the idea. The Standard newspaper played its part in applying the pressure:

'Darwin died, as he had lived, in the quiet retirement of the country home which he loved; and the sylvan scenes amidst which he found the simple plants and animals that enabled him to solve the great enigma of the Origin of Species may seem, perhaps, to many of his friends the fittest surroundings for his last resting place. But one who has brought such honour to the English name, and whose death is lamented throughout the civilized world…should not be laid in a comparatively obscure grave…we owe it to posterity to place his remains in Westminster Abbey, among the illustrious dead who make that noble fane unrivalled in the world.'[321]

So the funeral arrangements for Downe were cancelled. On Tuesday 25 April, three of Darwin's children followed their father's coffin as it travelled the sixteen rain-soaked miles from Downe to Westminster Abbey on a hearse drawn by four horses. The next day eight pallbearers gently picked up the wooden box, draped in black velvet, and carried it reverently to the centre of the Abbey transept. This was a moment steeped in symbolism and meaning more powerful than perhaps we can now imagine. Although the Queen and the Prime Minister were not present, thousands of dignitaries and ordinary people were. This was a nation informally mourning the loss of someone they knew was great. It was also a moment when the burial of the very man whose controversial ideas had torn society asunder became a statement that brought people back together.

The coffin was lowered into the grave dug a few feet from the monument to Isaac Newton, and the choir sang the same verse performed at the funeral of another British national hero, Lord Nelson; 'His body is buried in peace, but his name liveth evermore'. The eight pallbearers looked on, their job done. They may have turned to each other, perhaps shaken hands and shared their condolences and brief recollections. It made perfect sense for John, Huxley, Spottiswoode and Hooker to help carry their master to his final resting place. Importantly they were joined by George Campbell, the eighth Duke of Argyll, one of their archenemies in the heated debates of the 1860s (Desmond & Moore, 1991).[322] He still did not share Darwin's views but had grown to respect his intellectual brilliance and rightful place in the nation's pantheon of science.

Ironically, although Darwin was buried in Westminster Abbey as a national hero, he never received a knighthood from his sovereign; an honour he would have loved (Desmond & Moore, 1991). The proposal had been vetoed many years before by the Queen's Anglican advisors including

Bishop Wilberforce. Apparently only in death was Darwin safe, although Wilberforce undoubtedly turned in his grave when the doors of the Abbey were thrown open to his nemesis in 1882. Of the X Club members, only two received knighthoods earned in recognition of their individual work: Hooker and Frankland. Darwin's bulldog, Huxley, never received a knighthood in Britain despite the King of Sweden knighting him in 1873, along with Hooker and Tyndall. Spencer, Busk, Tyndall, Spottiswoode and Hirst received no public recognition from the British state. Evans and Prestwich, both of whom had never become centrally embroiled in the religious debate, were knighted in 1892 and 1896 respectively.

However, the fortunes of John were a little different. He had inherited his baronetcy in 1865 and on 29 December 1889 he received a letter from the Prime Minister, Lord Salisbury, informing him that the Queen 'has been pleased to direct that you should become a member of the Privy Council on the approaching New Years day'.[323] The Privy Council was a body of advisers to Queen Victoria who frequently met at Osborne House on the Isle of Wight. Two years after John was appointed, perhaps by coincidence, Huxley was also called upon to become a member of the Privy Council. On 9 July 1897, John received a further memo from Salisbury, informing him that the Queen had approved his membership of the Royal Commission being set up for the British Section of the Paris International Exhibition in 1900.

On 1 January 1900, John's postbag suddenly exploded with literally hundreds of letters from across the world: all congratulating him on the newspaper announcements in the *Times*. In the New Year's Honours List, 'Sir John Lubbock' had been elevated to the Peerage. Who in the audience at the 1860 BAAS meeting in Oxford might have guessed that forty years later one of Darwin's closest disciples would have the right to sit in the House of Lords; or that a daughter of Pitt Rivers would become a Lady? How proud and amused John's Master would undoubtedly have been.

One of the envelopes John opened with his paper knife contained a short note from his dear old friend, Hooker:

'Accept my own and my wife's such hearty congratulations on your long over-Earned recognition, & offer the same to Lady Lubbock.'[324]

Hooker, Spencer and Evans were the only brothers in arms still alive to share this moment with John. After twenty years of inspiring a major shift in society's view of science and religion, these comrades had begun their gradual walk into the sunset during the 1880s. A year after they had buried Darwin,

John, Huxley and Hooker were back in Westminster Abbey burying the first of their dear X Club companions. Spottiswoode died on 26 June 1883 from typhoid fever at the age of just fifty-eight. Only a month before, he had been at the Walsinghams for the evening with Pitt Rivers, John and Mrs Busk.

'Poor Spottiswoode died – a terrible loss.'[325]

When Busk died at the age of seventy-nine on 10 August 1886, the X Club was rocked but still a force to be reckoned with:

'X dinner at Athenaeum. Huxley, Hooker, Tyndall, Frankland, Hirst & I. Very cheery & quite like old times, though we much regretted poor Busk's absence.'[326]
'He has been a kind & good friend for many years, & his example was one to do everyone good. He was not only able but most good & kind. His friendship was certainly one of the greatest privileges of my life.'[327]

By the 1890s the X Club was a shadow of its former self. Hirst died in February 1892 and Tyndall passed away on 4 December 1893, aged seventy-three, after taking an accidental overdose of chloral hydrate that he used to treat his insomnia. Then the life of its founder and champion, Huxley, ended on 29 June 1895 in his seventieth year and Frankland died on 9 August 1899. Spencer had always been an independent character and was no different as he grew older. John, however, kept in close social touch with the Hookers and the families of Darwin and Busk throughout these later years.[328] Evans was also a close friend and was able to share in his namesake's joy at becoming a Peer of the Realm.

The critical question on everyone's minds was the title John would take for his peerage:

'I am longing to know what you will call yourself – I am half hoping you will take the name of Avebury.'[329]

On Tuesday 23 January 1900 he cleared out his locker at the House of Commons:

'It is sad to feel that my career there is over! Signed my dear old name for the last time.'[330]

The next day, at the age of sixty-six years, John picked up his pen and for the first time began signing his name as 'Lord Avebury'. So, this small village and its prehistoric stone circle in Wiltshire, which is today recognized as a World Heritage Site, became immortalized in the hereditary peerage of Great Britain.

During the last thirteen years of his life, as Lord Avebury, John continued in his efforts to champion the protection of ancient monuments from destruction. In the early months of 1901, the bastion of prehistory, Stonehenge itself, became the subject of contention. Sir Edmund Antrobus, the private owner of this most important monument, had asked for advice from the Society of Antiquaries regarding the safety of the stones. A committee was established, of which John was a member. There are indications that John offered funding at this point to support Stonehenge.[331] In the end no such transactions took place and a programme of remedial work and excavation was undertaken by the Society and paid for by Antrobus himself. However, the resolution passed by the Society also agreed to the erecting of a fence that would restrict public access to the site, a course of action that John could not countenance:

'I am much disturbed by the resolution passed this afternoon. The present right of way through Stonehenge gives the public a right of access. If however the roadway is diverted as proposed by the Committee, this right would be abandoned.

'We might I feel sure trust Sir Edmund, but if the suggestion is carried out, his successors might charge any fee they like for admittance or exclude the Public altogether.

'I cannot therefore but feel that the proposal is very dangerous & ought to be resisted in the interests of the public. At any rate, I cannot consent to be responsible for the suggestion & must ask you therefore to accept my resignation much as I regret to take such a course.'[332]

John was also only too aware of the imperfections in the ancient monuments legislation he had driven through Parliament in 1882. Pitt Rivers discussed these in length with John in a letter dated 30 August 1886.[333] For example, the 1882 Act gave no power of enforcement to the Inspector and Pitt Rivers found it unpleasant talking to private owners he did not know 'to ask them to do what they don't want to do.' Protection of prehistoric monuments in Ireland needed to be addressed, and he wanted to ensure that the government reimbursed Inspectorate expenses involved in travelling the country. It was also a disappointment that Roman and Medieval monuments were not included in the Act. By 1889, a network of local committees and correspondents had been created whose responsibility it was to monitor sites at a local level and converse with the landowners when required. John was president of the Kent committee.[334] The Ancient Monuments Protection Act (1882) Amendment (Ireland) was passed in 1892.

One of John's first actions in the House of Lords was to steer the passing of the Ancient Monuments Protection Act (1900) which extended certain powers of the 1882 Act. The notion of state guardianship was now applied to any monument identified of national importance, rights of public access were strengthened, and the power of purchase was extended to local authorities as well as the state. He continued to champion this interest in the Lords over the next ten years. Some modern commentators, with the benefit of hindsight, suggest that because these Acts were far from perfect, they should also be regarded as failures. Perhaps the 1882 Act did result in only a few English, Scottish and Welsh sites being placed under state guardianship, but this simple conclusion is missing the point. John and his friends placed the principle of ancient monuments protection firmly on the map. Less than twenty years after his death, three-thousand monuments were scheduled and protected.

It was the day after his seventieth birthday when John first took the Chair at the Society of Antiquaries of London in Burlington House. He had spent the night before with all his children gathered around the dinner table. Now he was surrounded by the great and good of London's antiquarian and archaeological community to mark his election as President of this illustrious society.[335] He held this office for the next four years, welcoming new members with a few words and a shake of the hand, whilst carrying the symbolic mace of office. He presided over lectures and new member elections where Fellows crowded around the ballot boxes waiting to cast their vote either for or against prospective candidates. He heard about new developments, ideas and excavations, listened to the new generation of archaeologists and scientists, and imparted his wisdom in the comfort of the Fellows' Room.

He continued his work as a British Museum Trustee and published new contributions on flowers, insects and geology. He worked on revised editions of familiar works, including *Pre-historic Times*, and received various awards and accolades. On Thursday 8 January 1903, Evans wrote to inform his old friend that the Geological Society had decided to award him the first Prestwich Medal. In the same year, he was awarded the Commander of

Legion of Honour by the French President. Three years earlier he had made the first Huxley Memorial Lecture at the Royal Anthropological Institute, and was the President of numerous societies.

John still, very occasionally, collected prehistoric stone implements and went on archaeological excursions. In 1898, he attended the BAAS in Bristol with Evans and Frank Darwin. During this trip, Mr Ralls gave him two recently discovered Palaeolithic axes from Broom Quarry near Axminster.[336] In the same year, he also bought a collection of prehistoric stone implements from Abydos with the help of the British Museum[337] and was gifted two axes found in Kent from an old local friend, Benjamin Harrison.[338] In 1902 he acquired a small collection of Palaeolithic axes from another recent discovery in the Clapton area of London, donated by the archaeologist studying the sites, Worthington Smith.[339]

By now, his travelling companions were primarily his children, and his motivation was as much about sharing his enthusiasm for prehistory with the next generation as visiting the sites themselves. On Monday 5 May 1899, he took his fourteen year-old daughter, Ursula, on a day out by train to Ightham in Kent:

'Mr. Harrison met us where he met with me 27 years ago! We went to the rock shelter, & to the knoll of hard rock at Oldbury. Then to see his collection. Then up to the summit of the plateau where there is a pit showing at least 25 feet of lower Tertiary gravels, like the Holwood but with more clay & especially chalk. Caught the 5.40 back.'[340]

In the same year, Benjamin Harrison gave John two Palaeolithic axes and a stone flake which were found at Ightham and Nosted. On Saturday 30 August 1907, John took his sons Eric and Maurice (aged fourteen and six respectively) to see Harrison:

'First we looked at his collection. Then to Terry's Hill, to the Reservoir Field, to Ash, up the dry Stansted valley, to the chalk escarpment from which we had a splendid view, then to the Wrotham gault pit & finally to the great lower Greensand pit. Found several eoliths. The patches of ochreous gravels are very curious & suggest old river beds.'[341]

The last two items ever recorded in the catalogue to John's collection were Palaeolithic implements from the Ightham Plateau, found at Ash and Farndale.[342] John's handwriting states that Mr Harrison presented them, but no date of acquisition is given. However, they came from the exact locations walked by John, Harrison, Eric and Maurice on that summer day in 1907. Now they sit in a cardboard box in the stores at Bromley Museum.

On Wednesday 18 September, John notes in his diary:

'[Eric] is a dear little fellow…He has been acting as my secretary, & we have been working on pollen together'.[343]

This diary entry mirrored life sixty years earlier, when a fourteen year-old John had caught the eye of Darwin. One month later, John and his family arranged for 'the names of my scientific friends' to be hand painted on to some of the doors in their grand seaside residence at Ramsgate, Kingsgate Castle. The familiar names of 'Darwin', 'Hooker', 'Huxley', 'Tyndall', 'Busk', 'Evans' and 'Grant Duff' were all remembered in this way. In June 1911, John embarked for the last time on the journey to Hooker's house in the small village of Sunningdale near Windsor Great Park:

'Went down to lunch with Sir J. Hooker. Found him very well, but as he will be 93 in a few days I fear it may be the last meeting of the X Club!'[344]

Six months later:

'Dear Old Hooker died: the last of the old X Club…a wonderful man & most kind friend. The last of our little group.'[345]

This final sentence was not quite correct; John was the last of that little group. He had been the younger of Darwin's first disciples and he outlived all of his original 'brothers in arms'. Pitt Rivers had died in 1900 and Evans passed away on 31 May 1908:

'Received the sad news of Sir John Evans' death. One of my oldest & greatest friends. It has been one of the privileges of my life to have enjoyed his friendship.'[346]

After Hooker's death, John became the only living connection to that heady intellectual world of Darwinism that had changed so many things. He kept his interest in science, prehistory and ethnography right until the end of his life. On Saturday 9 November 1912, he recorded his last archaeological excursion with a colleague at the British Museum to see flints from the base of the Cray Valley in Kent.

'They are most interesting. Those I had seen before did not convince me, but those now exhibited seem to me to be certainly worked intentionally.'[347]

On Saturday 22 February 1913 he attended the British Museum meeting where the 'Piltdown Man' remains were shown to the Trustees for the first time; 'the most simian of any yet found'.[348] They examined the fragments of a skull and jawbone allegedly collected in 1912 from a gravel pit at Piltdown in Sussex by Charles Dawson. The experts of the day thought they were looking at the fossilized remains of an early human species, but it was exposed as an elaborate hoax, forty years after John's death in 1953. He returned home to Kingsgate Castle to work on his new edition of *Pre-historic Times* and immediately added a section on Dawson's discovery, describing the skull as the 'most ape-like of all'. Which of course the forger, who had combined the skull of a modern man with the lower jawbone of an orang-utan, had probably intended!

On Thursday 1 April he recorded in his diary that he had mainly been working on *Pre-historic Times*, seeking to answer his critics. He completed the manuscript for the final time and sent it to his publishers, Williams & Norgate. However, he did not live to see this seventh edition of his great work published. John Lubbock, First Lord Avebury, died of heart failure as he returned to High Elms from Kingsgate Castle on 28 May 1913. His final words on prehistoric archaeology, ethnography and Darwin's theory of evolution were published in November of that year.

John is buried close to his chosen place of worship, the church of *St Giles the Abbott* in the village of Farnborough Kent, perched on the northern slope of the North Downs.[349] His grave was marked by a tall Celtic style stone cross, carved with symbols of some of his life's interests, including Bronze Age axes and a large stone megalith (Plate VIII).[350] The magnificent yew tree that watched over John's funeral procession as it filed past in early June 1913 still stands guard in the corner of the churchyard.

John left behind a wife, nine children, and a collection of prehistoric stone tools that hold the secrets about his remarkable contribution to Darwin's campaign for evolution and natural selection.

Chapter 9

Letting the Stones Speak

I opened the old brown corrugated cardboard box that felt cold and damp to the touch and smelt of must. A heap of screwed up newspapers lay bundled inside discoloured by time and nibbled by silverfish. In each crumpled sheet hid an ancient stone waiting to tell its story. I put my hand in to the box as if it were a lucky dip, gently cradled the first wrapped artefact I found inside and carefully lifted it out and on to the table.

The newspaper concealed a long, thin object that was heavier than I expected. I carefully unwrapped it to reveal a large, beautifully polished axe crafted out of a rich mustard-brown mottled flint (Plate III). The fine workmanship and skill of the Danish craftsmen thousands of years ago had created a true work of art.

Every artefact I found in the pile of old storage boxes was an object of wonder; however big or small, whether a few thousand or hundreds of thousands of years old. What were they? Who had made them and for what purpose? How had they been made so perfectly without the technologies we rely on so heavily today? Who was this John Lubbock, the first Lord Avebury, and why had he collected these objects? My fascination with prehistory started at the age of sixteen as a volunteer at my local museum in Bromley.

Written in black ink directly on to the giant Danish axe in John's very distinctive handwriting was the number '303'. I turned to the unassuming handwritten notebook that he had written, which lay open on the table to one side, and turned through the pages to find the catalogue entry for 303. It was recorded as a flint axe from 'Magleby Stevas. Seeland.' that John acquired in 1865. These were quite literally just foreign words to me at the time but I stored up the knowledge and wanted to find out more. For the time being I gave each artefact a unique museum reference number and repacked and stored them in new archival boxes. Ten years later I began my doctorate research into the collections that has formed the basis of this published remembrance to John and his prehistoric collection.

When John passed away in May 1913, his personal memories of the rich and detailed stories held within his collection died with him. We can now only pick through the clues that remain to help us uncover its meaning and importance in the past and the present. It is like piecing together a jigsaw with most of the pieces missing and the others scattered across the landscape; which is of course the fun and challenge of the historical and archaeological research that motivated John.

How did these collections survive for us to study and experience in the present? Some collectors of this time worked hard before their death to find a museum or similar organization to look after their collection once they had died, however, John does not appear to have done so. John made no explicit reference to his collection or Avebury in his last will and testament.[351] His 'museum' and archaeological property in Wiltshire are caught up anonymously in the general property and estate willed to his eldest son, John Birkbeck, the Second Lord Avebury. Perhaps this reflected his original belief that collections were tools for research and only of value when being used as such? Perhaps he believed that John Birkbeck and other members of his family were interested enough in the collection that they would continue to care for it? Perhaps he did not think the collection significant enough to preserve for the nation?

However, a man called Charles Hercules Read felt differently. In 1896, he had taken over the mantle from Franks as Keeper of the Department of British and Medieval Antiquities and Ethnography in the British Museum. In 1913 he was also the President of the Society of Antiquaries of London, having succeeded John in 1908 and so it is likely that John and Read knew each other well. As President of the Anthropological Institute in 1900, Read had written to John inviting him to give the first Huxley lecture in memory of his friend.[352] Read had presumably visited High Elms on more than one occasion to see the collection, given his knowledge of its contents, and on 27 May 1916 the following entry was made in the British Museum's Book of Presents:

'Sir Hercules Read has been familiar with the majority of these articles [Lubbock's collection] for many years past, and he took an opportunity recently to point out to the present Lord Avebury the importance of some of the series in the gift for the Museum collections. Eventually Lord Avebury offered as a gift a large series from his late father's collection. Among these by far the most important is a find from Hallstatt, in Upper Austria, excavated about 1863-9. It consists of a fine bronze bucket and a number of other articles in bronze and iron, illustrating the transition from bronze to iron in that part of the world... Such another series does not exist in England, and it will add greatly to the interest of the early section of the Early Iron Age Gallery. In addition to the foregoing Lord Avebury has also sent a large number of stone implements from various parts of the world, Egypt, France, America and the South Seas.

'The gift as a whole is of such importance that it deserves, in Sir Hercules' opinion, the special thanks of the Board.'[353]

The British Museum accession register records the generous donation of 356 artefacts from the second Lord Avebury in 1916. Most were from the Hallstatt excavations commissioned by John and Evans in the 1860s[354] including entry 1916.6-5.356:

'Bronze bucket with some holes, handles missing. With five broad ribs separating zones with rhombs of raised dots, the top and bottom zone with embossed ducks and wheel patterns, the base indented at centre. H. 12.'[355]

This bucket and other Hallstatt metal artefacts immediately became the subject of a British Museum publication, and nearly 100 years later still sit majestically on display in the prehistoric galleries of the British Museum. The stone implements are tucked away behind the scenes in drawers and boxes stored on endless steel racks. They wait for researchers like me to literally show them the light of day for their next fifteen minutes of fame, before they return to the darkness where they are protected. Safe, yet forgotten.

Read had the foresight to retain for the British Museum the type specimens figured in *Pre-historic Times*;[356] a very scientific course of action which suggests he at least regarded *Pre-historic Times* as an important work in the development of a discipline. He also arranged for other museums across the country to select material for their own collections:

'I can now report progress with the prehistoric collection you handed over to me some months ago.

'Having made my selection for the Museum, I had an announcement made at the meeting of the Museums Association in July with regard to the remaining part of the collection, and this announcement was printed in the August number of the "Museums Journal". The result was that I received applications from 29 museums, stating their desires. Of course there was overlapping in the demands, but all got something that they wanted, and I had regard in the distribution to the importance of the Museum.

'All are grateful, and I have great pleasure in conveying to you their thanks for the gifts. I am sure they will be useful, distributed as they are all over the country.'[357]

John Birkbeck Lubbock was fifty-five years old when he inherited the peerage on the death of the First Lord Avebury. In 1914, he oversaw the merger of the family bank with Coutts & Co., of which he became a partner. He was a member of the Board on several London, New Zealand and Australian banks, investment trusts and insurance companies. As the Second Lord Avebury he also continued to have an interest in the archaeological part of his father's life. He took over as a Trustee of the Christy Collection and wrote to the British Museum each year in this official capacity; offering as gifts to the British Museum Trustees all the objects acquired by the Christy Collection during the last year.[358]

He also kept a significant part of his father's collection and appears to have continued collecting himself. On 1 March 1921, he donated 'a miscellaneous collection of coins and medals' to the British Museum that might have been collected either by himself or his father. Three months later, he donated:

'Gold hoop with circular cupshaped ends, found near [Mullingar], Co. Westmeath, in a peat bog; a gold torc found in Norfolk; & two bracelets of twisted silver wire, probably Indian. Given by Lord Avebury, High Elms, Farnbro', Kent...C.H. Read.'[359]

This was one of the last acquisitions recorded by Read who retired in the same year. Lord Avebury was a subscriber who contributed financially to the illustrated tribute for Read which was published to mark this occasion.[360] It is possible that Avebury's donations to the British Museum during 1921 are also part of this tribute, as these and the original 1916 donations are the only three occasions when such personal gifts were made.

The Second Lord Avebury died in 1929, unmarried and with no children to inherit the title and estate meaning it passed to the fourteen year old son of Harold Fox Pitt Lubbock, the eldest son of the marriage between John and Alice. The Third Lord Avebury, also called John, sold the High Elms estate to Kent County Council in 1938 at the age of twenty-three, two years after he legally came of age and was able to manage his own inheritance. King's College Hospital London leased the mansion and transformed it into the King's College Hospital Preliminary Training School for nurses. In 1962 ownership was transferred to the Orpington Urban District Council and shortly afterwards to the London Borough of Bromley who designated the estate as a Green Belt Open Space. In 1967 the mansion was destroyed by fire, but the parkland remains open today for the public and keen golfers to enjoy.

Shortly after the Second World War, in 1947, the Third Lord Avebury was persuaded to give the remainder of his grandfather's prehistoric archaeology and ethnographic collection to a group of amateur historians based locally near High Elms.[361] These dedicated enthusiasts were led by Andrew Fordyce, a local road construction company owner, and Mr Eldridge, a council workman, in their production of exhibitions of local history and archaeology during 1945 and 1946.[362] Under the banner of the newly formed Orpington Historical Society, the group had organized the 1946 exhibition with a clear mission – to create a museum for Orpington. Lord Avebury donated his collections to their cause on the condition that they remained within the Orpington area.[363]

Having such an important historical collection could only strengthen the case for the development of a new museum, yet it took another eighteen years of lobbying and hard work before their dream became a reality.[364] Progress had looked very hopeful in 1957 when the Orpington Urban District Council (OUDC) took over responsibility for libraries, museums and art gallery provision in Orpington from Kent County Council.[365] Steps were immediately taken to identify the Grade II Listed Orpington Priory as the museum venue and budgets were allocated. However, nothing further happened until 1963 when the OUDC was spurred into action by the imminent transfer of their affairs to the London Borough of Bromley in 1965:

'Nothing happened until Orpington was about to be incorporated into Bromley, when the Council quickly decided to appoint a curator and establish a museum in the Priory.'[366]

LETTING THE STONES SPEAK

Mrs Adelaide Lubbock, mother of the Fourth Lord Avebury, opened the museum to great local fanfare. This new cultural facility was part of a wider programme to improve the village's civic amenities during 1963-5, including drawing up plans for a swimming pool and shopping centre as a last defiant act before Orpington became part of Bromley.[367] In the meantime, however, John's collections were homeless and since 1959 had been stored in completely inadequate conditions at St Mary's Cray Library. A selection of the 'best pieces' saw the light of day in 1965 and filled the Priory showcases with their stories, but many other specimens remained wrapped up in their boxes silently waiting to be given a new voice and purpose.

Over 100 miles away the stones of Avebury told a very different story. The land containing the stone circle had stayed in the ownership of the Lubbock family until 1934[368] when the whole village and stone circle was bought outright by a millionaire whose family had made their money from marmalade. Alexander Keiller was an enthusiastic archaeologist determined to excavate the Avebury circle and re-create the site. In the late 1930s he spent three summers working on a quadrant of the circle each year; clearing the undergrowth, uncovering the stones and re-erecting them where he felt they had probably originally been located. Where there were gaps because stones had been destroyed he inserted concrete posts to mark their presence. He invested £50,000 of his own money, but then the war came and all archaeological work stopped. In 1943 the National Trust bought the 950-acre site from Keiller for £12,000. By the mid-1960s, Avebury had become a great day out for local families on August Bank Holidays! In 1987, its international importance was confirmed when Avebury was designated a World Heritage Site by UNESCO, together with Stonehenge, Silbury Hill and the wider prehistoric landscape within which they sit.

Today, over 350,000 people visit Avebury each year and a few thousand people visit Bromley Museum to find out more about John and his collection. The stories of these places, people and prehistoric artefacts are closely intertwined yet no reference is made to their connection. Without this now anonymous collection of stone implements and ethnographic artefacts sitting in a small local museum in the outskirts of London, the future World Heritage Site at Avebury would probably not have survived into the twentieth century.

Stay quiet and put your ear to these stones at Avebury and Bromley. They whisper to you from an age long past about a man and his friends who fundamentally changed the way we view the world and simultaneously popularized the idea of prehistory.

Why does this story matter in a twenty-first century world where we share information, photographs, music and moving images across the world at the touch of a button; travel in cars, trains and airplanes to places once remote and distant but now only a few hours away; where we can cure cancer and eradicate smallpox; where many of us are entitled by law to an education up to the age of eighteen; and have an equal opportunity to apply for a job based on our individual abilities and experience.

It matters because by opening up the idea of prehistory, John and his fellow Darwinists contributed significantly to a shift in world thinking that makes our lives of today possible. It provided evidence to undermine the Ussher Chronology argument that the earth is a mere 6,000 years old, and replaced it with the idea of natural evolution. These brothers in arms finally undermined once and for all the basic premise of British Society that had been paramount until this point: that you needed to be a member of the established Anglican Church and under the patronage of the aristocracy to succeed.

The idea of prehistory has helped shape a world in which all of us in Britain can take freedom of thought and belief, a right to education and the ability to choose our own destiny for granted. It has contributed to the development of a science that has lost its shackles of conservative religious values and is free to explore the full richness of our universe. With DNA research we are able to find cures for fatal genetic illnesses; with the exploration of our environment we are able to recognize the threat of global warming and identify ways of reducing its impact. Through developing our understanding of prehistory over the last 150 years we have been able to demonstrate that all people across the world are part of one human family. New DNA research, with its roots in Darwinian science, provides strong evidence of our shared ancestry from a small group of early humans living in Africa millions of years ago.

Darwin's unique selling point and the reason why he was so successful in making the case for evolution was his absolute belief in evidence. It was so much harder to refute an idea if there were clear concrete physical examples based on evidence to swing the argument. The prehistoric sites that John protected and the collections of artefacts that he, Evans and others acquired were major pieces of evidence. They were both the keys that unlocked the door to prehistory and human evolution, and the vital ammunition in the debate that overpowered the opposition. Today, they sit anonymously in the landscape, or in the gallery display case and the museum store. In reality, they are incredibly important symbols of that seismic shift in our thinking. They are a reminder that the world is continuously shaped by the people who live in it, the ideas that they have and the courage it takes to share these ideas with others.

Despite the importance of having knowledge of prehistory there is currently no space for this topic in the National Curriculum, unless an innovative teacher introduces it creatively into another subject area.

'Prehistory does not feature in the national curriculum in England... the UK is the only European state to neglect prehistory in this way... Prehistory should be part of all national curricula.'

(All-Party Parliamentary Archaeology Group, 2003, p.28)

Our knowledge of our past begins 2,000 years ago with the Romans and the Saxons; even shorter than the Ussher Chronology that John, Darwin and many others fought so hard to challenge! Only those people who go on to study relevant subjects at university, or are interested in the Discovery Channel or National Geographic, ever really find out more.

How can we as individuals and a society truly know ourselves and shape our future lives if we do not understand where we sit in the bigger picture? Over the last 150 years we have been in the privileged position of realizing that we are all minor mortal cogs in a large cosmological wheel. A moment in a continuum of time that stretches back millions of years and, as long as we look after the earth whilst we are here, will stretch forward way into the future. Yet we take it so much for granted that we ignore it. We forget that we are very temporary caretakers in this timeless story. We ignore our distant ancestors, all that they achieved and all that they handed down to us – in our genes, our values and beliefs, our country and its landscape. By hiding prehistory from ourselves and from our children we deny a large part of who we are and what we will pass on to future generations.

Sitting today in the Red Lion pub, built in the heart of the Avebury stone circle over 400 years ago, it is easy to see that some people are curious about this enigmatic prehistoric landscape. The dining room is heaving with young families, leather-clad bikers, classic car enthusiasts and National Trust members; all having Sunday lunch before they stroll around the stones and visit the museum and shops. Groups of foreign tourists have their cameras at the ready, taking photographs of black-coated pagan worshippers and ley-liners that wander through this landscape with a different sense of purpose and belonging.

These visitors may be looking for different answers but their questions are often the same. Where did the stones come from? Why are they there? Who built the circle? How were they moved to this location and erected as a circle? Why is there a ditch around the circle? Why is there a village built in the middle of it? How long ago did this all take place? They enjoy the green rolling scenery and tranquil calm of this protective place, and navigate their way through the minefield of sheep droppings and molehills to get up close and touch the rough and misshaped sarsen stones with their hands. They imagine they can see faces and the shapes of animals and dinosaurs in the bumps, cracks and curves of these massive eroded boulders. Some have tied coloured ribbons on tree branches; while young children run down the banks of the dyke, followed closely by their parents shouting about the sea of stinging nettles waiting for them at the bottom.

Many may part with their money to visit the museum and find out more about the site they are visiting. Postcards, guidebooks, fossils and birthstones are bought to take back home as souvenirs. Yet few visitors leave this precious site at Avebury with a real sense of the relevance it has to our modern lives. This is because we as a society have forgotten about the importance of prehistory, why it matters to all of us, and why reconnecting Darwin, evolution, John Lubbock and prehistory is so important to the future ambitions of our human race.

This is John's true legacy to us, but it is only his stones that hold the depths of this secret. Let them speak to you and listen to what they say.

Bibliography

All-Party Parliamentary Archaeology Group (2003) *The Current State of Archaeology in the United Kingdom: First Report of the All-Party Parliamentary Archaeology Group*. Kent: Caxton & Holmesdale Press.

Anonymous (1945) Orpington's story. *Orpington & Kentish Times*, 21 September.

Bangert, S. (2008) Evans, Scandinavia and international exchange networks. In: A. MacGregor (ed.) (2008) *Sir John Evans 1823-1908: Antiquity, Commerce and Natural Science in the Age of Darwin*. Oxford: Ashmolean Museum, pp. 231-256.

Barton, R. (1976) *The X Club: Science, Religion, and Social Change in Victorian England*. Ph.D, University of Pennsylvania, unpublished.

Beechey, F.W. (1831) *Narrative of a Voyage to the Pacific and Beering's Strait, to Co-operate with the Polar Expeditions*. London: Henry Colburn and Richard Bentley.

Belcher, E. (1861) On the manufacture of works of art by the Esquimaux. *Transactions of the Ethnological Society of London*, New Series 1, pp.129-45. London.

Bell, J. and Balbis, A. (1832) *A System of Geography, Popular and Scientific, Or, A Physical, Political, and Statistical Account of the World and its Various Divisions*. Archibald Fullarton.

Bockstoce, J.R. (1977) *Eskimos of Northwest Alaska in the Early Nineteenth Century*. Oxford: University of Oxford.

Bowden, M. (1991) *Pitt Rivers: The Life and Archaeological Work of Lieutenant-General Augustus Henry Lane Fox Pitt Rivers DCL, FRS, FSA*. Cambridge: Cambridge University Press.

Bowler, P. (1989) *Evolution: The History of an Idea.* Revised edition. Berkeley: University of California Press.

Bowler, P. (1993) *Biology and Social Thought 1850-1914.* Berkeley: Office for History of Science and Technology, University of California.

Burkhardt, F. (ed.) (2008) *Charles Darwin: The 'Beagle' Letters.* Cambridge: Cambridge University Press.

Burkhardt, F. & Smith, S. (eds.) (1988) *The Correspondence of Charles Darwin, Volume 5: 1851-1855.* Cambridge: Cambridge University Press.

Burkhardt, F. & Smith, S. (eds.) (1990) *The Correspondence of Charles Darwin. Volume 6: 1856-1857.* Cambridge: Cambridge University Press.

Burkhardt, F. & Smith, S. (eds.) (1991) *The Correspondence of Charles Darwin, Volume 7: 1858-1859.* Cambridge: Cambridge University Press.

Burkhardt, F. *et al.* (eds.) (1993) *The Correspondence of Charles Darwin, Volume 8: 1860.* Cambridge: Cambridge University Press.

Burkhardt, F. *et al.* (eds.) (1994) *The Correspondence of Charles Darwin, Volume 9: 1861.* Cambridge: Cambridge University Press.

Burkhardt, F. *et al.* (eds.) (1997) *The Correspondence of Charles Darwin, Volume 10: 1862.* Cambridge: Cambridge University Press.

Burkhardt, F. *et al.* (eds.) (1999) *The Correspondence of Charles Darwin, Volume 11: 1863.* Cambridge: Cambridge University Press.

Burkhardt, F. *et al.* (eds.) (2001) *The Correspondence of Charles Darwin, Volume 12: 1864.* Cambridge: Cambridge University Press.

Burkhardt, F. *et al.* (eds.) (2003) *The Correspondence of Charles Darwin, Volume 13: 1865.* Cambridge: Cambridge University Press.

Carlyle, T. (1849) Occasional discourse on the Negro Question. *Fraser's Magazine.* London.

BIBLIOGRAPHY

Chapman, W. (1989) The organisational context in the history of archaeology: Pitt Rivers and other British archaeologists in the 1860s. *Antiquaries Journal*, 69, pp.23-42. London: Society of Antiquaries.

Chippindale, C. (1990) *Who Owns Stonehenge?* London: Batsford.

Cook, J. (1997) A curator's curator: Franks and the Stone Age collections. In: M. Caygill and J. Cherry (eds.) (1997) *A.W. Franks. Nineteenth Century Collecting and the British Museum*. London: British Museum Press, pp.115-29.

Cooper, M. (1999) *Bryce McMurdo Wright Father and Son, 50 Years of Mineral Dealing, 1845-95*. Unpublished.

Crouch, W. (1904) *Bryan King and the Riots at St. George's-in-the-East*. London: Methuen.

Darwin, C. (1842) *The Structure and Distribution of Coral Reefs*. London: Smith, Elder & Co.

Darwin, C. (1851-4) *A Monograph on the Sub-Class Cirripedia*, Vol. 1 and 2. London: Ray Society.

Darwin, C. (1859) *On the Origin of Species*. London: John Murray.

Darwin, C. (1871) *The Descent of Man and Selection in Relation to Sex*. London: John Murray.

Desmond, A. (1994) *Huxley: The Devil's Disciple*. London: Michael Joseph.

Desmond, A. and Moore, J. (1991) *Darwin*. London: Michael Joseph.

Evans, J. (1943) *Time and Chance. The Story of Arthur Evans and his Forebears*. London: Longmans, Green & Co.

Evans, J. (1867) On some discoveries of stone implements in Lough Neagh, Ireland. *Archaeologia*, Vol. 41, pp.397-408. London: Society of Antiquaries of London.

Evans, J. (1872) *The Ancient Stone Implements, Weapons and Ornaments of Great Britain*. London: Longmans & Co.

Fitzroy, R. (1839) *Narrative of the Surveying Voyages of HMS's Adventure & Beagle between the years 1826 & 1836,* Volume 2. London: Henry Colburn.

Franklin, J. (1828) *Narrative of a Second Expedition to Shores of Polar Seas in years 1825, 1826 and 1827*. London: John Murray.

Gamble, C. and Kruszynski, R. (2009) John Evans, Joseph Prestwich and the stone that shattered the time barrier. *Antiquity*, 83, pp.461-475. York.

Gash, N. (1978) After Waterloo: British society and the legacy of the Napoleonic Wars. *Royal Historical Society Transactions*, 5(28), pp.145-57. London: Royal Historical Society.

Gough, B. (ed.) (1973) *To the Pacific and Arctic with Beechey: The Journal of Lieutenant George Peard of HMS "Blossom" 1825-8* Cambridge: Cambridge University Press.

Grant Duff, A. (ed.) (1924) *The Life-Work of Lord Avebury (Sir John Lubbock) 1834-1913*. London: Watts & Co.

Hall, N.J. and Burgis, N. (1983) *Trollope, Anthony. The Letters.* Stanford: Stanford University Press.

Healey, E. (2001) *Emma Darwin: The Inspirational Wife of a Genius.* London: Review Books.

Hermansen, V. (1934) *J.J.A. Worsaae. Af en Oldgrandskers Breve 1848-1885*. Copenhagen.

Hutchinson, H. (1914) *Life of Sir John Lubbock Lord Avebury*. London: Macmillan and Co.

Huxley, T. (1858) On the agamic reproduction and morphology of Aphis – Part I and II. In: *Transactions of the Linnean Society of London,* p.193-236. London: Linnean Society.

BIBLIOGRAPHY

Huxley, T.H. (1863) *Man's Place in Nature*. London: Macmillan.

Jensen, J.V. (1991) *Thomas Henry Huxley. Communicating for Science*. London: Associated University Press.

[Ker, D.], (1880) Along the Gotha Canal. *The New York Times*, July 14.

King, J. (1997) In M. Caygill and J. Cherry (eds.) (1997) *A.W. Franks. Nineteenth Century Collecting and the British Museum*. London: British Museum Press, pp.136-59.

Kipling, R. (1899) The White Man's Burden. *McClure's Magazine*. New York: S.S. McClure Co.

Lubbock, E. (1864) The ancient shell-mounds of Denmark. In F. Galton (ed.) *Vacation Tourists and Notes of Travel in 1862-3*. London: Macmillan.

Lubbock, J. (1853a) On Labidocera, Description of a new genus of Calanidae, and two new subgenera of Calanidae. *Annals and Magazine of Natural History*, XI, pp.25,202. London: Taylor and Francis Ltd.

Lubbock, J. (1853b) On two new species of Calanidae, with observations on spermatic tubes of Pontella Diaptomus etc. *Annals and Magazine of Natural History*, XII, pp.115,159. London: Taylor and Francis Ltd.

Lubbock, J. (1854) On some Arctic species of Calanidae. *Annals and Magazine of Natural History*, XIV, p.125. London: Taylor and Francis Ltd.

Lubbock, J. (1855a) On the freshwater entomostraca of South America. *Transactions of the Royal Entomological Society of London*, 8, pp.232-240. London.

Lubbock, J. (1855b) On the objects of a collection of insects. In Stainton, H.T. (ed.) *The Entomologists' Annual for MDCCCLV*. London: John Van.

Lubbock, J. (1856) On some entomostraca collected by Dr. Sutherland, in the Atlantic Ocean. *Transactions of the Royal Entomological Society of London*, 9, pp.8-37. London.

Lubbock, J. (1857a) An account of the two methods of reproduction in Daphnia, and of the structure of the Ephippium. *Philosophical Transactions of the Royal Society*, 147, pp.79-100. London: Royal Society.

Lubbock, J. (1858) On the digestive and nervous system of Coccus hesperidum. *Proceedings of the Royal Society*, IX, p.480. London: Royal Society.

Lubbock, J. (1859) On the ova and pseudova of insects. *Philosophical Transactions of the Royal Society*, 149, pp. 341-369. London: Royal Society.

Lubbock, J. (1861) On the køkkenmøddings: recent geologico-archaeological researches in Denmark. *Natural History Review N.S.* 1(IV), pp.489-504. London: Williams & Norgate.

Lubbock, J. (1862a) Evidence of antiquity of Man, afforded by Somme Valley. *Natural History Review N.S.* 2(VII) July, pp.244-269. London: Williams & Norgate.

Lubbock, J. (1862b) The ancient lake-habitations of Switzerland. *Natural History Review N.S.* 2(V), pp.26-51. London: Williams & Norgate.

Lubbock, J. (1863a) Review of 'The Antiquity of Man From Geological Evidences: with Remarks on the Theories of the Origin of Species and Variation' by Sir Charles Lyell. *Natural History Review N.S.* 3(X), pp.211-219. London: Williams & Norgate.

Lubbock, J. (1863b) A visit to the ancient shell-mounds of Scotland. *Natural History Review N.S.* 3(XI), pp. 415-422. London: Williams & Norgate.

Lubbock, J. (1864) Cave-men. *Natural History Review N.S.* 4(XV), pp.407-428. London: Williams & Norgate.

Lubbock, John (1865) *Pre-historic Times*. First edition. London: Williams & Norgate.

Lubbock, J. (1869) *Pre-historic Times*. Second edition. London: Williams & Norgate.

BIBLIOGRAPHY

Lubbock, J. (1870) *On the Origin of Civilisation and Primitive Condition of Man* First edition. London: Longmans & Co.

Lubbock, J. (1892) *The Beauties of Nature, and the Wonders of the World We Live In*. London: Macmillan & Co.

Lubbock, J. (1902) *On the Origin of Civilisation and Primitive Condition of Man* Sixth edition. London: Longmans & Co.

Lubbock, J. (1911) *Marriage, Totemism and Religion. An Answer to Critics*. London: Longmans & Co.

Lubbock, J. (1913) *Prehistoric Times,* Seventh edition. London: Williams & Norgate.

Lubbock, J.W. (1833) *On the Theory of the Moon and on the Perturbations of the Planets*. London.

Lubbock, J.W. (1835) *An Elementary Treatise on the Computation of Eclipses and Occultations*. London.

Lubbock, J.W. (1838) *Remarks on the Classification of the Different Branches of Human Knowledge*. London.

Lubbock, J.W. (1839) *An Elementary Treatise on the Tide*s. London.

Lubbock, J.W. (1861) *The Discovery of the Planet Neptune. Lecture delivered at the Bromley Literary Institute*. London.

Lyell, C. (1863) *Geological Evidences of the Antiquity of Man*. London: J. M. Dent & Son.

MacGregor, A. (2008) Sir John Evans, Model Victorian, Polymath and Collector. *Sir John Evans 1823-1908: Antiquity, Commerce and Natural Science in the Age of Darwin*. Oxford: Ashmolean Museum, pp. 1-33

Mayhew, H. (1861) *London Labour and the London Poor*, III. London: Griffin, Bohn & Co.

McClintock, A. (1995) *Imperial Leather: Race, Gender and Sexuality in the Colonial Contest.* London: Routledge.

Mckenzie, K. (2005) *Scandal in the Colonies: Sydney and Cape Town 1820-50.* Melbourne: Melbourne University Press.

Merrington, P. (1995) Pageantry and primitivism: Dorothea Fairbridge and the aesthetics of union. *Journal of South African Studies*, 21(4), pp.643-56. London: Taylor & Francis.

Milner, R. and Lane, J. (2009) Seeing corals with the eye of reason. *Natural History*, February 1. New York: American Museum of Natural History.

Morris, M. (1996) *Towards a sociology of Bronze Age studies in England.* M. Phil. Thesis, University of Durham, unpublished.

Murray, T. (2009) Illustrating 'savagery': Sir John Lubbock and Ernest Griset. *Antiquity, 83*, pp.488-499. York.

Nilsson, S. (1868) *The Primitive Inhabitants of Scandinavia.* London: Longmans, Green & Co.

Owen, J (1999) The collections of Sir John Lubbock, the first Lord Avebury (1834-1913): an open book? *Journal of Material Culture volume 4(3),* pp.283-302. London: Sage.

Owen, J. (2000) The collecting activities of Sir John Lubbock (1834-1913). Ph.D, University of Durham, unpublished.

Owen, J. (2006) Collecting artefacts, acquiring empire. *Journal of the History of Collections* 18(1), pp.9-25. Oxford: Oxford University Press.

Patton, M. (2007) *Science, Politics & Business in the Work of Sir John Lubbock: a Man of Universal Mind.* Aldershot: Ashgate.

Porter, B. (1987) *Britain, Europe and the World 1850-1986: Delusions of Grandeur.* London: Allen & Unwin.

BIBLIOGRAPHY

Seely & Paget (Chartered Architects) (1957) Report concerning the preservation and adaptation of the Priory to provide library and ancillary facilities. Unpublished.

Sherratt, A. (1983) John Evans and archaeology in the nineteenth century. Unpublished.

Sillitoe, P. (2005) The role of section H at the British Association for the Advancement Of Science. *Durham Anthropology Journal volume 13(2)*. Durham: Durham University.

Stocking, G. (1987) *Victorian Anthropology*. New York: Free Press, Macmillan.

Tylor, E.B. (1865) *Researches into the Early History of Mankind and the Development of Civilisation*. London: John Murray.

Tylor, E.B. (1871) *Primitive Culture: Researches into the development of mythology, philosophy, religion, language, art and custom*. London: John Murray.

Tyndall, J. (1883) *Hours of Exercise in the Alps*. New York: Appleton and Company.

Van Riper, B. (1993) *Men Among the Mammoths: Victorian Science and the Discovery of Human Prehistory*. Chicago: University of Chicago Press.

Wiell, S. (1996) A letter from Line: the Flensburg antiquities and the Danish-Prussian/ Austrian War of 1864. *Antiquity*, 70(270), pp.913-920. York.

Primary Archive and Collection Sources

The following primary archives and collections have been consulted during the course of researching and writing this book:

The Lord Avebury Collection, housed at the British Museum and at the London Borough of Bromley Museum, Orpington, Kent

The Archives of Wilhelm Boye, housed at the National Museum Copenhagen

The Correspondence of Charles Darwin, with the kind permission of the Darwin Correspondence Project, the Darwin family and Cambridge University Press

The Archives of Sir John Evans, housed at the Ashmolean Museum Oxford

The Archives of Sir Joseph Dalton Hooker, housed at the Royal Botanical Gardens Kew

The Archives of Sir John Lubbock, housed at the British Library and remaining with the Lubbock family, with the kind permission of the Lubbock family

The Archives of Japetus Steenstrup, housed at the Royal Library Copenhagen

Notes

Introduction
1. British Library MS Add 62683 114. "John Lubbock Diary entry dated 19 October 1892".
2. Alternative spelling 'Down'.
3. The china used by the family would have been Spode given as a wedding present and used for large gatherings. (Lyulph Lubbock Personal Communication, June 2012).
4. British Library MS Add 49655 64-65. "Letter from Augustus Franks to John Lubbock, 13th September 1890". British Library MS Add 49655 71-72. "Letter from Augustus Franks to John Lubbock, 18th September 1890".

Chapter 1
5. Darwin also gave John an arrowhead from Scotland in 1864. Bromley Museum, "Avebury Catalogue Volume 1", no. 267.
6. Figure 156 in the first edition of *Pre-historic Times* 'represents the head of a Fuegian harpoon, which closely resembles the ancient Danish specimen figured in page 80'. (Lubbock, 1865, p.436).
7. John William Lubbock published various papers regarding astronomical matters (including Lubbock, J.W., 1833; 1835; 1838; 1839; 1861).
8. The existence of a 'General Notebook on Natural History 1848 –' written by John and held in the archives of Down House, provides a tantalizing insight into this early relationship. It contains a diary of various observations, including sketches and a brief discussion of the 'races of men' on page 93. English Heritage, Down House Collections 88203395.
9. Down House Collections 88203395.
10. British Museum of Natural History Palaeontology Archive "Letter from Charles Darwin to G. Waterhouse [1850]".
11. "Letter from Charles Darwin to George Newport, 24 July [1851]" (Burkhardt & Smith, 1988, p.51).
12. "Letter from Charles Darwin to George Newport, 12 August [1851]" (Burkhardt & Smith, 1988, p.54-5).
13. British Library MS Add 62679. "John Lubbock Diary entry dated 22 February 1853".

14. British Library MS Add 62679. "John Lubbock Diary entry dated 22 September 1853".
15. British Library MS Add 62679. "John Lubbock Diary entries dated 30 November and 15 December 1853".
16. "Letter from Charles Darwin to J.D. Dana, 27 September [1853]" (Burkhardt & Smith, 1988, p.157).
17. British Library MS Add 62679. "John Lubbock Diary entry dated 27 October 1854".
18. British Library MS Add 62679. "John Lubbock Diary entry dated 5 February 1855".
19. British Library MS Add 62679. "John Lubbock Diary entry dated 9 March 1855".
20. British Library MS Add 62679. "John Lubbock Diary entry dated 5 September 1855".
21. "Letter from Charles Darwin to John Lubbock, [14 January 1856]" (Burkhardt & Smith, 1990, p.20).
22. British Library MS Add 62679. "John Lubbock Diary entry dated 27 June 1855".
23. "Letter from Charles Darwin to John Lubbock, 19 July 1855" (Burkhardt & Smith, 1988, p.382).
24. British Library MS Add 62679. "John Lubbock Diary entry dated 9 December 1855".
25. "Letter from Charles Darwin to John Lubbock, 24 April 1856" (Burkhardt & Smith, 1990, p.87).
26. "Letter from Charles Darwin to Sir J.W. Lubbock, 27 May [1856]" (Burkhardt & Smith, 1990, p.114).
27. "Letter from Charles Darwin to John Lubbock, 27 October [1856]" (Burkhardt & Smith, 1990, p.250-1); "Letter from Charles Darwin to John Lubbock, 1 November [1856]" (ibid., p.254-5).
28. "Letter from Charles Darwin to John Lubbock, 11 August [1857]" (Burkhardt & Smith, 1990, p.442).
29. "Letter from Charles Darwin to John Lubbock, [22 November 1857]" (Burkhardt & Smith, 1990, p.489-90). John was elected a fellow in 1858.
30. "Letter from Charles Darwin to John Lubbock, 14 July [1857]" (Burkhardt & Smith, 1990, p.430).
31. Parthenogenesis is a form of asexual reproduction found in females, where growth and development of embryos occurs without fertilization by a male (Wikipedia, 6th April 2011).

32. British Library MS Add 62679. "John Lubbock Diary entry dated 3 January 1858"; Lubbock, J. (1858).

33. George Busk was a professor at the Royal College of Surgeons.

34. "Letter from Charles Darwin to John Lubbock, 30 [March 1858]" (Burkhardt & Smith, 1991, p.58).

35. "Letter from John Lubbock to Charles Darwin, 10 June 1858" (Burkhardt & Smith, 1991, p.105).

36. "Letter from Charles Darwin to John Lubbock, [August-September 1858]" (Burkhardt & Smith, 1991, p.143).

37. "Letter from Charles Darwin to John Lubbock, [November 1858]" (Burkhardt & Smith, 1991, p.179).

38. "Letter from Charles Darwin to John Lubbock, [6 February 1859]" (Burkhardt & Smith, 1991, p.244).

39. "Letter from John Lubbock to Charles Darwin, 8 February 1859" (Burkhardt & Smith, 1991, p.245).

40. "Letter from Charles Darwin to John Lubbock, 9 February [1859]" (Burkhardt & Smith, 1991, p.246).

41. "Letter from Charles Darwin to John Lubbock, 17 December [1859]" (Burkhardt & Smith, 1991, p.436). John admired an important book in the High Elms Library, published in 1844 and entitled *Vestiges of the Natural History of Creation* (Lyulph Lubbock, personal communication, June 2012). This controversial but popular book, originally published anonymously, is widely credited as preparing the way for *Origin* by beginning to raise questions about transmutation and its relationship to religious views.

42. "Letter from Charles Darwin to John Lubbock, 8 March [1859]" (Burkhardt & Smith, 1991, p.257).

43 "Letter from John Lubbock to Charles Darwin, 15 March 1859" (Burkhardt & Smith, 1991, p.266).

44. "Letter from Charles Darwin to John Lubbock,16 March [1859]" (Burkhardt & Smith, 1991, p.266-7).

45. "Letter from Charles Darwin to John Lubbock, 21 [March 1859]" (Burkhardt & Smith, 1991, p.267).

46. "Letter from Charles Darwin to Alfred Russel Wallace, 6 April 1859" (Burkhardt & Smith, 1991, p.279).

47. "Letter from Charles Darwin to John Lubbock, 14 December [1859]" (Burkhardt & Smith, 1991, p.432-3).

48. "Letter from Charles Darwin to John Lubbock, 17 December [1859]" (Burkhardt & Smith, 1991, p.436).

49. "Letter from Charles Darwin to Joseph Dalton Hooker, 22 [June 1859]" (Burkhardt & Smith, 1991, p.308).

Chapter 2

50. Sources for this description are Mayhew (1861, pp.302-4) and 'Warehouses. From the Grey River' by the artist and engraver, Mortimer Menpes, created 1889 (National Maritime Museum reference no. PU8069).

51. [Ker, D.], (1880, p.3); British Library MS Add 62679. "John Lubbock Diary entry dated 7 September 1863".

52. "Letter from Charles Darwin to Joseph D. Hooker, 3 March 1860" (Burkhardt et al, 1993, p.116).

53. The accounts shared with Darwin vary slightly in detail between the various players - Huxley, Hooker and John - as can be seen in the series of letters exchanged between these four individuals immediately afterwards (Burkhardt et al, 1993, pp.279-80). Lubbock's contribution was in response to Wilberforce and Fitzroy stating that 'the embryology of the individual in many cases represents the past history of the species'. University of Cambridge, DAR 106/7 (series 4): 30 "Letter from John Lubbock to Francis Darwin, 2 January 1896".

54. British Library MS Add 49639 30-54. Various correspondence about efforts to support the Essayists, including Dr Frederick Temple.

55. "Letter from John Lubbock to Charles Darwin, 23 August 1862" (Burkhardt et al, 1997, p.381).

56. "Letter from J.J.A. Worsaae to his wife, dated 18 July 1863" (Hermansen, 1934). Translated for me by Anne Katrine Gjerløff.

57. "Letter from Japetus Steenstrup to Charles Darwin, 8 April 1852" (Burkhardt & Smith, 1991, p.487-9).

58. Copenhagen Royal Library NKS 3460 to. "Letters from John Lubbock to Japetus Steenstrup, 9 July, 7 November and [July-November] 1861".

59. Copenhagen Royal Library NKS 3460 to. "Letter from John Lubbock to Japetus Steenstrup, 7 Nov 1861".

60. Copenhagen Royal Library NKS 3460 to. "Letter from John Lubbock to Japetus Steenstrup, 24 March 1862".

61. Copenhagen Royal Library NKS 3460 to. "Letter from John Lubbock to R. Puggaard, 17 April 1862".

62. British Library MS Add 49640 48. "Letter from Japetus Steenstrup to John Lubbock, [1]0 March 1863".

63. "Letter from Charles Darwin to Asa Gray, 23 February 1863" (Burkhardt et al, 1999, p.166-7).
64. Copenhagen Royal Library NKS 3460 to. "Letter from John Lubbock to Japetus Steenstrup, [February 1863]".
65. Copenhagen Royal Library NKS 3460 to. "Draft letter from Japetus Steenstrup to John Lubbock, undated [1863]"; British Library MS Add 49640 71-2 "Letter from Japetus Steenstrup to John Lubbock, 21 June 1863".
66. Copenhagen National Museum arkiv IV 118 Boye. Original catalogue, handwritten in Danish, of Wilhelm Boye's collection.
67. Bromley Museum, "Avebury Catalogue Volume 1", nos. 188-204.
68. Bromley Museum, "Avebury Catalogue Volume 1", nos. 230-242; Census information 1860 and 1870 held at the Danish State Archives, Copenhagen.
69. Bromley Museum, "Avebury Catalogue Volume 1", nos. 205-214.
70. Copenhagen Royal Library NKS 3460 to. "Letter from John Lubbock to Japetus Steenstrup, 12 September 1863".
71. Bromley Museum, "Avebury Catalogue Volume 1", no. 252.

Chapter 3

72. Ashmolean Museum, John Evans Archive: Notebook 1866, Austria, England, France; 1867, France, Switzerland, Germany, England; 1868, England, France. Entries dated 14-24 April 1866.
73. Ashmolean Museum, John Evans Archive: Notebook 1866, Austria, England, France; 1867, France, Switzerland, Germany, England; 1868, England, France. Entries dated 14-24 April 1866.
74. Bromley Museum "Avebury Catalogue Volume 1", nos.361-2, 366-7.
75. Ashmolean Museum, John Evans Archive. "Letters between John Evans and Joseph Stapf 1866-1876". Transcribed by Dr. Andrew Sherratt.
76. Ashmolean Museum, John Evans Archive. "Letters between John Evans and Joseph Stapf 1866-1876". Transcribed by Dr. Andrew Sherratt.
77. Ashmolean Museum, John Evans Archive. "Letters between John Evans and Joseph Stapf 1866-1876". Transcribed by Dr. Andrew Sherratt.
78. Ashmolean Museum, John Evans Archive. "Letters between John Evans and Joseph Stapf 1866-1876". Transcribed by Dr. Andrew Sherratt.
79. Bromley Museum, "Avebury Catalogue Volume 1", nos. 255-262 and 266.
80. Bromley Museum, "Avebury Catalogue Volume 1", nos. 268-273.

81. Bromley Museum, "Avebury Catalogue Volume 1", nos. 290-293.
82. "Letter from Charles Darwin to Thomas Huxley 5 July 1860" (Burkhardt et al, 1993, p.280).
83. "Letter from John Lubbock to Charles Darwin, 7th [February] 1862" and "Letter from Charles Darwin to William Erasmus Darwin, 14 February [1862]" (Burkhardt et al, 1997, p.73 and 79-80).
84. "Letter from Joseph Dalton Hooker to Charles Darwin, [10th March 1862]" Burkhardt et al, 1997, p.104-5).
85. "Letter from Charles Darwin to Joseph Dalton Hooker, 14th March [1862]" Burkhardt et al, 1997, p.115).
86. British Library MS Add 49642 167-170 "Letter from John Tyndall to John Lubbock, 3 September 1868" describing a picture of John.
87. British Library MS Add 62680. "John Lubbock Diary entry dated Christmas 1864".
88. "Letter from Charles Darwin to Charles Lyell, 4 May 1860" (Burkhardt et al, 1993, pp.188-9).
89. "Letter from Charles Darwin to Charles Lyell, 4 May 1860" (Burkhardt et al, 1993, pp.188-9).
90. "Letter from Charles Darwin to Charles Lyell, 8 [May 1860]" (Burkhardt et al, 1993, p.196).
91. "Letter from Charles Darwin to John Lubbock, 14 August [1861]" (Burkhardt et al, 1994, p.235).
92. "Letter from Ellen Lubbock to Emma Darwin, [January 1862]" (Burkhardt et al, 1997, p.1).
93. Bromley Museum "Avebury Catalogue Volume 1", no. 267, arrow from Scotland.
94. The story of this harpoon '277' is a future research project in its own right. In 1842, Darwin also donated two Tierra del Fuego harpoons to the museum of his Cambridge tutor, Joseph Henslow, in Ipswich. Were these the Jemmy Button harpoons? Did Darwin collect other harpoons during the *Beagle* voyage from the Fuegian communities he encountered, one of which he gave to John? He would have had opportunity to do so. Bromley Museum 81.21.1 is a fragment of a barbed harpoon (20cm long) that is a possible candidate for '277', although it is not confirmed.
95. Bromley Museum, "Avebury Catalogue Volume 1", nos. 265; 285-9; 294; 296-8.
96. Bromley Museum, "Avebury Catalogue Volume 1", no. 299.
97. Bromley Museum, "Avebury Catalogue Volume 1", nos. 279-284; 295; 300.

Chapter 4

98. Today called the Rupununi River.

99. Royal Botanical Gardens Kew Archives, Volume 204, 225-229 "Letter from WH Campbell to JD Hooker, 9 December 1867".

100. Royal Botanical Gardens Kew Archive, Volume 204, 230 "Letter from WH Campbell to JD Hooker, 23 December 1867".

101. Bromley Museum, "Avebury Catalogue Volume 2", no. 580.

102. British Library MS Add 62680 "John Lubbock Diary entry 23 June 1872".

103. In June 1886, for example, John commented in his diary on how well attended the X Club was for a change, with Huxley, Tyndall, Frankland, Hirst and him being present, but he also remarked on the fact that Busk could not be there for reasons of ill health. Busk died on 10 August 1886. British Library MS Add 62683 "John Lubbock Diary entry 3 June 1886". See also footnote 104.

104. An exchange of correspondence between Huxley and Hooker during 1883 reflects a rivalry emerging within the X Club membership. Spottiswoode had died whilst in office as President of the Royal Society. John had sought the prize but concerns over his ability to commit adequate time to the role resulted in Huxley being elected; 'the only intimate friend who is absolutely silent is Lubbock. So I suppose he thought the pear was for him'. Royal Botanical Gardens Kew Archive "Letters from TH Huxley to JD Hooker 1854-95 158-163".

105. British Library MS Add 49640 "Letter from Thomas Huxley to John Lubbock, 2 May [1863]".

106. See Chapter 2.

107. Ashmolean Museum, John Evans Archive: letters between Evans, John, Pitt Rivers and Prestwich regarding the Anthropological Institute elections of December 1872.

108. Hence the successive presidents of the Royal Anthropological Institute were John (1871-3), Busk (1873-4), Pitt Rivers (1876-7), Evans (1877-9), and Pitt Rivers again (1881-2).

109. Bromley Museum, "Avebury Catalogue Volume 1", nos. 328-331; Ashmolean Museum, John Evans Archive: "Notebook 1863, France, England etc.: 1864 with Prestwich, France, England: 1865, France, England, Denmark?, Germany: 1866, France" Entries for July 3 1864 and June 11th 1865.

110. British Library MS Add 49641 47-8 "Letter from John Evans to John Lubbock, 16 June 1865".

111. Bromley Museum, "Avebury Catalogue Volume 1", nos. 429-439.
112. Bromley Museum, "Avebury Catalogue Volume 2", no. 799.
113. Bromley Museum, "Avebury Catalogue Volume 2", no. 1187.
114. Ashmolean Museum, John Evans Archive: "Notebook 1866, Austria, England, France; 1867, France, Switzerland, Germany, England; 1868, England, France".
115. Bromley Museum, "Avebury Catalogue Volume 1", nos. 355-9.
116. Ashmolean Museum, John Evans Archive: "Notebook: 1872, France Gratz, Agram: 1873, coins: 1874, France, Denmark: 1875, France: 1876, France, Italy, Austria: 1877, France, England".
117. Bromley Museum, "Avebury Catalogue Volume 2", no. 1115.
118. Ashmolean Museum accession numbers 1927.1396; 1927.1429-1430.
119. Ashmolean Museum accession numbers 1927.6148a-c.
120. Bromley Museum "Avebury Catalogue Volume 2", no. 1091.
121. Royal Botanical Gardens Kew Archive: Extract from the Instructions for collecting plants and seeds for the Garden and the Herbarium, and the useful products of vegetables. Letter from Lesley Price, Archivist, Royal Botanical Gardens, Kew to Janet Owen dated 11 July 1997.
122. Bromley Museum, "Avebury Catalogue Volume 1", nos. 458-9.
123. Bromley Museum, "Avebury Catalogue Volume 1", no. 576.
124. Bromley Museum, "Avebury Catalogue Volume 1", no. 506.
125. Bromley Museum, "Avebury Catalogue Volume 1", no. 575.
126. Bromley Museum, "Avebury Catalogue Volume 2", nos. 580 and 594.
127. Bromley Museum, "Avebury Catalogue Volume 2", nos. 686-7, 737 and 1046.
128. Bromley Museum, "Avebury Catalogue Volume 2", no. 981.
129. Bromley Museum, "Avebury Catalogue Volume 2", no. 1029.
130. Bromley Museum, "Avebury Catalogue Volume 2", no. 880.
131. Bromley Museum, "Avebury Catalogue Volume 2", no. 1111.
132. Bromley Museum, "Avebury Catalogue Volume 2", nos. 1074, 717 and 898.
133. Bromley Museum, "Avebury Catalogue Volume 2", nos. 731, 801-2, 581 and 614.

Chapter 5
134. See chapter 3.
135. Bromley Museum, "Avebury Catalogue Volume 1", no. 258.
136. British Museum accession number 1916.6-5.4.

137. Copenhagen Royal Library NKS 3460 to. "Letter from John Lubbock to Japetus Steenstrup, [February 1863]".

138. "Letter from John Lubbock to Charles Darwin, 7 April 1863" (Burkhardt et al, 1999, p.298).

139. Copenhagen Royal Library NKS 3460 to. "Letter from John Lubbock to Japetus Steenstrup, 12 September 1863".

140. Copenhagen Royal Library NKS 3460 to. "Letter from John Lubbock to Japetus Steenstrup, 23 December 1863".

141. Copenhagen Royal Library NKS 3460 to. "Letter from Japetus Steenstrup to John Lubbock, 29 May 1864".

142. "Letter from Charles Darwin to John Lubbock, 11 June 1865" University of Cambridge Darwin Correspondence Project Letter 4858. Also published in (Burkhardt et al, 2003).

143. "Letter from Charles Darwin to John Lubbock, 11 June 1865". University of Cambridge Darwin Correspondence Project Letter 4858. Also published in (Burkhardt et al, 2003).

144. British Library MS Add 62680 "John Lubbock Diary entry dated December 1868".

145. See chapter 4.

146. See chapter 3; Royal Society of London Archives LUA 12 "John Lubbock accounts".

147. "Letter from Ellen Lubbock to Charles Darwin, [27 August - 1 September 1865]". University of Cambridge Darwin Correspondence Project Letter 4603. Also published in (Burkhardt et al, 2003).

148. British Library MS Add 49641 "Letter from Canon Greenwell to John Lubbock, 11 August 1866".

149. Bromley Museum, "Avebury Catalogue Volume 1", nos. 440-442.

150. Bromley Museum, "Avebury Catalogue Volume 1", nos. 480-2, 485-491.

151. Bromley Museum, "Avebury Catalogue Volume 1", nos. 554-5.

152. British Library MS Add 62680 "John Lubbock Diary entry 9 April 1868".

153. Bromley Museum, "Avebury Catalogue Volume 2", nos. 649-51.

154. Bromley Museum, "Avebury Catalogue Volume 2", nos. 749-62, 765-8.

155. British Library MS Add 49677 9-10 "A Catalogue of a valuable and highly interesting ARCTIC COLLECTION, made by the late Mr. Shingleton, during the voyages of the *Investigator*, *Enterprise* and *Fox* in search of Sir John Franklin".

156. Bromley Museum, "Avebury Catalogue Volume 1", no. 416.

157. Bromley Museum, "Avebury Catalogue Volume 2", nos. 769-71.

158. Bromley Museum, "Avebury Catalogue Volume 2", no. 786.

159. Jill Cook, Curator, British Museum, Personal communication, 1999.

160. Bromley Museum, "Avebury Catalogue Volume 1", nos. 578-9.

161. Bromley Museum, "Avebury Catalogue Volume 2", nos. 598-612.

162. The Wilhelm Boye collection was the first acquisition of a significant collection from Denmark, in 1863. See chapter 2.

163. Bromley Museum, "Avebury Catalogue Volume 1", no. 549.

164. National Museum Copenhagen. Privatsamlinger – Index Laegl: 1853-1910 "Letter from J.C.L. Petersen to M. F. Herbst, 9 July 1887".

165. Interestingly, the Avebury Catalogue entry immediately prior to the Petersen Collection acquisition (549) records acquisition of an 'Esquimaux scoop made of Musk Ox horn, Shingleton Collection. Pres. By Mr. Flower'. Bromley Museum, "Avebury Catalogue Volume 1", no. 548.

166. British Library MS Add 49677 "Letter from Conrad Engelhardt to [John Lubbock], 22 May 1867".

167. Bromley Museum, "Avebury Catalogue Volume 2", no. 643.

168. British Library MS Add 49677 "Letter from George Augustus Robinson to John Lubbock, 29 January 1866".

169. Bromley Museum, "Avebury Catalogue Volume 1", nos. 353, 407-415.

170. Bromley Museum, "Avebury Catalogue Volume 2", no. 740.

171. British Library MS Add 49677 "Letter from B. Plant to John Lubbock, 5 March 1868"; Bromley Museum, "Avebury Catalogue Volume 2", nos. 595-7.

172. Bromley Museum, "Avebury Catalogue Volume 1", nos. 476-7; Bromley Museum, "Avebury Catalogue Volume 2", no. 699.

173. Bromley Museum, "Avebury Catalogue Volume 1", no. 478.

174. British Library MS Add 49677 "Note from Walter Elliot to John Lubbock, 29 June 1869" (written on 'Travellers Club' headed notepaper); Bromley Museum, "Avebury Catalogue Volume 2", no. 780.

175. Bromley Museum, "Avebury Catalogue Volume 1", no. 564; Bromley Museum, "Avebury Catalogue Volume 2", no. 636.

176. Bromley Museum, "Avebury Catalogue Volume 1", nos. 443-450.

177. Bromley Museum, "Avebury Catalogue Volume 1", nos. 510-16, 518-21.

178. British Library MS Add 49677 "Letter from Chas C Abbott to John Lubbock, 6 June 1871".
179. British Library MS Add 49677 "Letter from Chas C Abbott to John Lubbock, 8 June 1871".
180. "Letter from Charles Darwin to Asa Gray, 15 August [1865]". University of Cambridge Darwin Correspondence Project Letter 4882. Also published in (Burkhardt et al, 2003).
181. British Library MS Add 62680 "John Lubbock Diary entry dated September 1870".

Chapter 6

182. We first encountered John Franklin in chapter 5 when our story connected with Franklin's final and fateful journey into the Arctic in search of the North West Passage. The expedition to which I now refer took place twenty years earlier and represents the voyage when Franklin came closest to discovering a route from the Atlantic to Pacific across the top of the world.
183. Bromley Museum, "Avebury Catalogue Volume 1", nos. 460-2.
184. British Library MS Add 49677 "Letter from Sir Edward Belcher to Sir John Lubbock, 17 June [1867]".
185. Bromley Museum, "Avebury Catalogue Volume 2", nos. 1004-1015.
186. Bromley Museum, "Avebury Catalogue Volume 1", Greenland antiquities a-z.
187. See chapter 5.
188. Bromley Museum, "Avebury Catalogue Volume 1", no. 299; Bromley Museum, "Avebury Catalogue Volume 2", no. 840.
189. British Library MS Add 49677 "Letter from J. Hector to John Lubbock, 27 April 1870".
190. Bromley Museum, "Avebury Catalogue Volume 2", no. 881.
191. British Library MS Add 49677 "Letter from Alfred Tozer to John Lubbock, 19 July 1872".
192. British Library MS Add 49677 "Letter from Harry Cecil Cameron to John Lubbock, 27 November 1874". The Channel Islands New Zealand Bound website records a Harry Cecil Cameron, born Manchester 29 November 1848, son of Charles Cameron and Julia Buckley. Left in 1862 on HMS *Conway* merchant training ship which was wrecked. He became a sheep farmer in New Zealand and returned to England in 1873.

193. Bromley Museum, "Avebury Catalogue Volume 2", no. 1080.
194. *Australian Dictionary of Biography* (Online Edition, 2009). Entry for Brazier, John William (1842-1930).
195. British Library MS Add 49677 "Letter from J. Brazier to John Lubbock, 11 May 1877"; Bromley Museum, "Avebury Catalogue Volume 2", nos. 1140-2.
196. British Library MS Add 49677 "Letter from J. Brazier to John Lubbock, 20 January 1880"; Bromley Museum, "Avebury Catalogue Volume 2", nos. 1169-1171.
197. British Library MS Add 49677 "Letter from J. Brazier to John Lubbock, [1883]"; Bromley Museum, "Avebury Catalogue Volume 2", no. 1195.
198. British Library MS Add 49677 "Letter from J. Brazier to John Lubbock, 28 April 1886".
199. Bromley Museum, "Avebury Catalogue Volume 2", nos. 1127-1134.
200. Bromley Museum, "Avebury Catalogue Volume 2", no. 1160.
201. Encyclopaedia Britannica Eleventh Edition (1910-11); Wikipedia 'Botocudo', 15th February 2009.
202. British Library MS Add 49677 "Letter from Chas Browne to John Lubbock, [1872]". See also chapter 4 re Christy Collection.
203. British Library MS Add 49677 "Letter from Bartle Frere to John Lubbock, 21 August 1874".
204. British Library MS Add 49677 "Letter from William Merewether to John Lubbock, 15 March 1875".
205. Bromley Museum, "Avebury Catalogue Volume 2", nos.1091-2.
206. British Library MS Add 49677 "Letter from Lady C. Frere to John Lubbock, 20 November 1877".
207. Website – Desmond McAllister;s Collaborated Genealogy, http://www.desmond-mcallister.info/Eng/Ancestors, 15th February 2009.
208. British Library MS Add 49677 "Letter from [Chas] A. Fairbridge to John Lubbock, 15 September 1873".
209. Bromley Museum, "Avebury Catalogue Volume 2", no. 1069.
210. Bromley Museum, "Avebury Catalogue Volume 2", nos. 888-897.
211. British Library MS Add 49677 33-34 "Letter from Julius Kessler to John Lubbock, 24 November 1870".
212. British Library MS Add 49677 "Letter from Andrew Swanzy to John Lubbock, 12 September [1870]" and "Letter from Winwoode Read to John Lubbock, 17 September 1870".

213. Bromley Museum, "Avebury Catalogue Volume 2", no. 841.
214. See chapter 1.
215. Royal Society of London Archives LUA5 "John Lubbock Notebook on Archaeology and Travels 1".
216. In September 1864, John borrowed the first and third volumes of the *Narrative of the Surveying Voyages of His Majesty's Ships Adventure and Beagle (Narrative)*; volume 1 covered the first expedition, from 1826 to 1830, and volume 3 was Charles Darwin's Journal and Remarks. University of Cambridge Darwin Correspondence Project Letter 4606. "Letter from John Lubbock to Charles Darwin, 2 September 1864". Also published in (Burkhardt et al., 2001).
217. Royal Society of London Archives LUA15 "John Lubbock Notebook 'Notes on Savages'".
218. Royal Society of London Archives LUA3 "John Lubbock Notebook 'Notes on Civilisation'".
219. See chapter 1.
220. A London-based artist with a shop in Suffolk Street, off Leicester Square, renowned for his studies of animals and satirical caricature.
221. For further information about the Griset paintings, please refer to Murray (2009). John also commissioned a twentieth illustration in 1871 with a very different subject matter. It depicts a Pacific atoll that is of small size and where the entire circle has been converted to land with the exception of a small inlet. Darwin in his 1842 publication entitled *The Structure and Distribution of Coral Reefs* regards these types of atoll as comparatively rare. He reproduced an image from Beechey (1831) of Whitsunday Island to illustrate this type. Maps reproduced to accompany the *Beagle Narrative* (Fitzroy, 1839) show North Keeling Island that also appears very similar to the subject of the Griset painting. The painting, now housed at Bromley Museum, also contains a small image of a ship that could represent HMS *Beagle*. It has been suggested by Milner & Lane (2009) that, given the subject matter, John commissioned it as a gift to Darwin that was never given.
222. Both Carlyle and Mill were visitors to High Elms as well as other critics of Darwinism. It is important to remember that there was a social dimension to John's personal life and that discussion with people holding differing views was an integral part of healthy intellectual debate.
223. This more empathetic view is perhaps illustrated by the 'Eyre Affair

of 1864' which is outlined in more detail in Patton, 2007. Governor Eyre of Jamaica brutally suppressed a rebellion of black peasants on the island and executed George Gordon, a mixed race local politician who was suspected of involvement. John Stuart Mill and a group of liberal intellectuals, including John, Darwin and other members of the X Club sought to prosecute Eyre in Britain for the murder of Gordon. Thomas Carlyle championed Eyre's defence and other supporters included Kingsley and Charles Dickens. The case never came to trial, but the debate provides an interesting insight into the political values of Darwinism.

Chapter 7

224. Bromley Museum "Avebury Catalogue Volume 2", nos.1149-1159.
225. British Library MS Add 49655 22-3 "Letter from Lord Derby to John Lubbock, 17 July 1890".
226. "Letter from Charles Darwin to Joseph Dalton Hooker, [29 July 1865]" University of Cambridge Darwin Correspondence Project Letter 4874. Also published in (Burkhardt et al, 2003).
227. See chapter 5.
228. Present-day Istanbul.
229. Bromley Museum, "Avebury Catalogue Volume 2", nos. 1034-5.
230. Bromley Museum, "Avebury Catalogue Volume 2", no.1031.
231. British Library MS Add 62680 "John Lubbock Diary entry dated September – October 1872".
232. British Library MS Add 62680 "John Lubbock Diary entry dated 20 November 1873".
233. British Library MS Add 62680 "John Lubbock Diary entry dated Wednesday 12th February 1873".
234. Bromley Museum, "Avebury Catalogue Volume 2", no.1077.
235. British Museum Unregistered material acquired from John Lubbock into the Christy Collection, listed and drawn in the British Museum Christy Catalogue 'Egypt Eg.1-37 + numbers and others': Eg.18-28 flint flakes presented by Sir John Lubbock 1874 (Thebes 20 November 1873); Eg. 29-36 flint flakes presented by Sir John Lubbock 1874 (Abydos 22 November 1873).
236. Bromley Museum, "Avebury Catalogue Volume 2", no.1109.
237. Bromley Museum, "Avebury Catalogue Volume 2", no.1108.
238. British Library MS Add 62680 "John Lubbock Diary entries dated 15 September – 13 October 1876".

239. British Library MS Add 62680 "John Lubbock Diary entry dated 4 June 1877".

240. Bromley Museum, "Avebury Catalogue Volume 2", no.1138.

241. British Library MS Add 62680 "John Lubbock Diary entry dated 3 December 1877".

242. British Library MS Add 62680 "John Lubbock Diary entries 3-24 December 1877".

243. British Library MS Add 62680 "John Lubbock Diary entry dated 24 December 1877".

244. Bromley Museum, "Avebury Catalogue Volume 2", nos. 1143-1147.

245. "Letter from John Lubbock to his mother, Harriet, 10 July 1865". Reprinted in Hutchinson 1914, p.77-8.

246. John and Nelly had one other child, Florence, who died a few hours after birth. (Lyulph Lubbock, Personal Communication, June 2012).

247. Bromley Museum, "Avebury Catalogue Volume 2".

248. British Museum accession number 1916. 6-5.32.

249. Bromley Museum, "Avebury Catalogue Volume 2".

250. British Library MS Add 62680 "John Lubbock Diary entry dated 13 November 1880".

251. See chapter 2.

252. "Letter from Charles Darwin to John Lubbock, 2nd August 1881" reprinted in Hutchinson, H., 1914, p.176.

253. For example, British Library MS Add 62683 "John Lubbock Diary entry dated 22 July 1883" describing a flower-spotting walk with Joseph Hooker at Hind Head; British Library MS Add 62683 "John Lubbock Diary entry dated 5 June 1889" describing a trip to Switzerland with Huxleys involving plant collecting and flower-spotting.

254. Contemporary account in *The News of the World* reprinted in Hutchinson, 1914, p.123-4.

255. British Library MS Add 62680 "John Lubbock Diary entry dated 4 July 1877" records when the University of London Senate decided to ask for a Charter enabling it to 'open all our degrees to women'.

256. British Library MS Add 62683 "John Lubbock Diary entry dated 14 July 1883".

257. British Library MS Add 62680 "John Lubbock Diary entry dated 12 March 1873".

258. British Library MS Add 62680 "John Lubbock Diary entry dated 25 June 1875".

259. British Library MS Add 62680 "John Lubbock Diary entries dated 10-12 March 1877".
260. British Library MS Add 62683 "John Lubbock Diary entries dated 10 and 11 October 1883".
261. British Library MS Add 62683 "John Lubbock Diary entries dated 8 and 9 June 1885".
262. British Library MS Add 62680 "John Lubbock Diary entries dated 4 January 1873 and 25 February 1874".
263. British Library MS Add 62680 "John Lubbock Diary entries dated 8 - 11 April 1874".
264. British Library MS Add 62680 "John Lubbock Diary entries dated 3 October and 7 December 1874".
265. British Library MS Add 62680 "John Lubbock Diary entry dated 2 December 1876".
266. British Library MS Add 62680 "John Lubbock Diary entry dated 15 February 1877".
267. For example, British Library MS Add 62683 "John Lubbock Diary entry dated 7 May 1883"; British Library MS Add 62683 "John Lubbock Diary entry dated 3 December 1885"; British Library MS Add 62683 "John Lubbock Diary entry dated 5 October 1888" 'Hooker & Hirst only'.
268. See footnotes 103-4.
269. British Library MS Add 62683 "John Lubbock Diary entry dated 18 November 1883".
270. British Library MS Add 62683 "John Lubbock Diary entry dated 9 April 1884".
271. British Library MS Add 62683 "John Lubbock Diary entry dated 7 May 1884".
272. British Library MS Add 62683 "John Lubbock Diary entry dated [9-16 April] 1884".
273. British Library MS Add 62683 "John Lubbock Diary entry dated 5 July 1884".
274. In 1884 Pitt Rivers donated his collection of prehistoric archaeology and ethnography to the University of Oxford.
275. British Library MS Add 62683 "John Lubbock Diary entries dated 12 January, 19 October and 16 November 1891, and 24 October 1892".
276. Photograph of delegates from the Bromley & District Teaching Association standing on the steps of High Elms with John dated 3 September 1898 (in Lubbock Family Archive).

277. British Library MS Add 62684 "John Lubbock Diary entry dated 28 September 1912".

278. See chapter 6.

279. Lyulph Lubbock, Personal Communication, 2010.

280. Royal Society of London Archives L3. "Notebook Book of Life: Letter from Napoleon III to John Lubbock, 10 April 1872".

281. British Library MS Add 62680 "John Lubbock Diary entry dated 18 February 1873".

282. British Library MS Add 49655 64-5 "Letter from A. W. Franks to John Lubbock, 13 September 1890".

283. British Library MS Add 62684 "John Lubbock Diary entry dated 7 July 1900".

284. Bromley Museum, "Avebury Catalogue Volume 2".

285. Bromley Museum, "Avebury Catalogue Volume 2".

286. British Library MS Add 62683 "John Lubbock Diary entry dated 24 September 1886".

Chapter 8

287. British Library MS Add 62680 "John Lubbock Diary entries dated 26 March - 1 April 1872".

288. Extract from Miss Sewell's *Experience of Life*, reprinted in Crouch, W. (1904), p.148

289. British Library MS Add 62680 "John Lubbock Diary entry dated Tuesday 30 June 1871".

290. British Library MS Add 49643 172-3 "Letter from B. King to John Lubbock, 3 October [1871]".

291. British Library MS Add 49643 174-8 "Letters between T. Kemms and John Lubbock, 27, 28 and 31 October [1871]".

292. Royal Society of London Archives LUA 12 "John Lubbock accounts".

293. British Library MS Add 49643 180-2 "Letter from B. King to John Lubbock, 18 November [1871]".

294. British Library MS Add 62680 "John Lubbock Diary entry dated Friday 29 March 1872".

295. British Library MS Add 49644 1 "Letter from A.C. Smith to John Lubbock, [4] January 1872".

296. British Library MS Add 49644 14-15 "Letter from Joseph Burtt to John Lubbock, 2 April 1872".

297. British Library MS Add 62680 "John Lubbock Diary entry in year summary for 1872".

298. Letter by James Fergusson published in *The Athenaeum* 13 Jan 1866.

299. Letter by John Lubbock published in *The Athenaeum* 15 Jan 1866.

300. British Library MS Add 49641 126 "Letter from T. Wright to John Lubbock, 17 April 1866".

301. British Library MS Add 49642 9-10 "Letter from A.C. Smith to John Lubbock, 23 Feb 1867".

302. British Library MS Add 49644 57-8 "Letter from A.C. Smith to John Lubbock, 19 June 1873".

303. British Library MS Add 62680 "John Lubbock Diary entry dated 28 June 1873" (entry later crossed out).

304. British Library MS Add 62680 "John Lubbock Diary entry dated 26 June 1875".

305. British Library MS Add 62680 "John Lubbock Diary entry dated 24 August 1876".

306. British Library MS Add 62680 "John Lubbock Diary entry dated 27 March 1872".

307. Royal Library Copenhagen NKS 3460 "Letter from John Lubbock to Japetus Steenstrup, [Feb 1863]".

308. British Library MS Add 49644 29-30 "Letter from John Lubbock to O. Morgan, 22 November 1872".

309. Hansard's Parliamentary Debates Third Series Vol. CCXV 24 March – 15 May 1873. London, Cornelius Buck, columns 1607-1608 (8 May 1873).

310. Hansard's Parliamentary Debates Third Series Vol. CCXVIII 5 March – 8 May 1874. London, Cornelius Buck, columns 574-595 HC (15 April 1874). The Second Reading was 'put off' for six months.

311. British Library MS Add 62680 "John Lubbock Diary entry dated 15 April 1874".

312. Hansard's Parliamentary Debates Third Series Vol. CCXVIII 5 March – 8 May 1874. London, Cornelius Buck, columns 574-595 HC (15 April 1874).

313. Hansard's Parliamentary Debates Third Series Vol. CCXXIII 18 March 1875 – 3 May 1875. London, Cornelius Buck, columns 879-917 (14 April 1875).

314. Hansard's Parliamentary Debates Third Series Vol. CCXXV 16 June 1875 – 23 July 1875. London, Cornelius Buck, columns 90-1 (17 June 1875).

315. Hansard's Parliamentary Debates Third Series Vol. CCL 5 February

1880 – 28 February 1880. London, Cornelius Buck, columns 370-2 (9 February 1880).

316. British Library MS Add 62680 "John Lubbock Diary entry dated 24 February 1880".

317. British Library MS Add 49645 192 "Letter from John Lubbock to A. Pitt Rivers, 25 October 1882".

318. British Library MS Add 62680 "John Lubbock Diary entry dated 19 March 1882".

319. British Library MS Add 62680 "John Lubbock Diary entry dated 20 April 1882".

320. British Library MS Add 49645 124-5 "Letter from John Lubbock to the Dean of Westminster, 21 April 1882".

321. Editorial published in *The Standard* 22 April 1882, reprinted in Desmond & Moore, 1991, pp.667-8.

322. See chapter 2.

323. British Library MS Add 49653 132-3 "Letter from Lord Salisbury to John Lubbock, 29 December 1889".

324. British Library MS Add 49666 100 "Letter from J. Hooker to John Lubbock, 2 January 1900".

325. British Library MS Add 62683 "John Lubbock Diary entry dated 27 June 1883".

326. British Library MS Add 62683 "John Lubbock Diary entry dated 3 June 1886".

327. British Library MS Add 62683 "John Lubbock Diary entry dated 10 August 1886".

328. British Library MS Add 62684.

329. British Library MS Add 49666 124-5 "Letter from H.H. Robinson to John Lubbock, 3 January 1900".

330. British Library MS Add 62684 "John Lubbock Diary entry dated 23 January 1900".

331. British Library MS Add 49668 72-3 "Letter from C. H. Read to John Lubbock, 9 March 1901".

332. British Library MS Add 49668 89 "Letter from John Lubbock to Mr. Dillon, 26 March [1901]".

333. British Library MS Add 49649 97-100 "Letter from A. Pitt Rivers to John Lubbock, 30 August 1886".

334. British Library MS Add 49653 97-8 "Letter from A. Pitt Rivers to John Lubbock, 20 October 1889".

335. British Library MS Add 62684 "John Lubbock Diary entries dated 27 and 28 April 1904".

336. Bromley Museum, "Avebury Catalogue Volume 2", nos.1171-2.

337. Bromley Museum, "Avebury Catalogue Volume 2", no.1168.

338. Bromley Museum, "Avebury Catalogue Volume 2", nos.1167 and 1169.

339. Bromley Museum, "Avebury Catalogue Volume 2", nos.1176-1178.

340. British Library MS Add 62684 "John Lubbock Diary entry dated 5 May 1899".

341. British Library MS Add 62684 "John Lubbock Diary entry dated 30 August 1907".

342. Bromley Museum, "Avebury Catalogue Volume 2", nos.1183-1184.

343. British Library MS Add 62684 "John Lubbock Diary entry dated 18 September 1907". Eric Fox Pitt Lubbock died at Ypres on the battlefield only ten years later.

344. British Library MS Add 62684 "John Lubbock Diary entry dated 21 June 1911".

345. British Library MS Add 62684 "John Lubbock Diary entry dated 10 December 1911".

346. British Library MS Add 62684 "John Lubbock Diary entry dated 31 May 1908".

347. British Library MS Add 62684 "John Lubbock Diary entry dated 9 November 1912".

348. British Library MS Add 62684 "John Lubbock Diary entry dated 22 February 1913".

349. John moved his place of worship to St. Giles, a couple of miles walk from High Elms, after the vicar at his regular church, St. Mary the Virgin at Downe, presented a sermon that condemned Darwin and his theories of evolution.

350. This memorial has subsequently been moved to a different location in the churchyard of St Giles the Abbott.

Chapter 9

351. The last Will and Testament of the Right Honourable John Baron Avebury of Avebury in the County of Wiltshire of Lombard Street in the City of London and of High Elms in the County of Kent. 22nd January 1913.

352. British Library MS Add 49667 66-7 "Letter from C.H.Read to John Lubbock, 4 April 1900".

353. British Museum Archive. "Book of Presents 1916 entry dated 27 May 1916".
354. See chapter 3.
355. British Museum accession register entry 1916.6-5.356.
356. British Museum Archive. "Letter from C.H.Read to Lord Avebury, 31 August 1916".
357. Ibid. These museums included those at Leeds, Nottingham, the Petrie Museum, Derby, Ipswich and Peterborough for example.
358. British Museum Archive. "Letter from C.H. Read to Trustees of the British Museum, 17 January 1921".
359. British Museum Archive. "Extract from the British Museum Report of Donations, Department of British and Medieval Antiquities, 3 June 1921".
360. British Museum Department of British & Medieval Antiquities (1921) "Charles Hercules Read: A tribute on his retirement from the British Museum to C.H. Read". Unpublished. The bronze bucket from Hallstatt collected by John Lubbock in 1869 and donated to the British Museum in May 1916 by the second Lord Avebury is featured in this tribute.
361. Orpington Historical Society Minute 1947.
362. Anonymous, 1945; Andrew Fordyce (first Chair of the Orpington Historical Society), Personal Communication, Spring 1990.
363. Orpington Historical Society Minute 1947.
364. Orpington Historical Society Minutes 25 May 1948, 3 September 1951 and 29 May 1956.
365. Orpington Urban District Council Minutes 24:751; 24:1111; 1545; Seely & Paget,1957; Andrew Fordyce, Personal Communication, Spring 1990.
366. "Letter written by Andrew Fordyce to the Editor, Orpington & Kentish Times [January 1974]" Published in the *Orpington & Kentish Times*, 24 January 1974.
367. Orpington Urban District Council Minutes 30:811.
368. Silbury Hill remains in the ownership of the Lubbock family, under guardianship of English Heritage.

Index

INDEX

INDEX

INDEX

Tracing Your Family History?

Read Your Family HISTORY

ESSENTIAL ADVICE FROM THE EXPERTS

Your Family History is the only magazine that is put together by expert genealogists. Our editorial team, led by Dr Nick Barratt, is passionate about family history, and our networks of specialists are here to give essential advice, helping readers to find their ancestors and solve those difficult questions.

In each issue we feature a **Beginner's Guide** covering the basics for those just getting started, a **How To …** section to help you to dig deeper into your family tree and the opportunity to **Ask The Experts** about your tricky research problems. We also include a **Spotlight** on a different county each month and a **What's On** guide to the best family history courses and events, plus much more.

Receive a free copy of *Your Family History* magazine and gain essential advice and all the latest news. To request a free copy of a recent back issue, simply e-mail your name and address to marketing@your-familyhistory.com or call 01226 734302*.

Your Family History is in all good newsagents and also available on subscription for six or twelve issues. For more details on how to take out a subscription, call 01778 392013 or visit **www.your-familyhistory.co.uk**.

Alternatively read issue 31 online completely free using this QR code

*Free copy is restricted to one per household and available while stocks last.

www.your-familyhistory.com